Golem

The Broom of the Czechs

The Decay of Czech Nationalism

Golem

The Broom of the Czechs

The Decay of Czech Nationalism

Walter Jacobi

Translated by Kytka Hilmarová

Golem - The Broom of the Czechs: The Decay of Czech Nationalism.
Copyright © 2023 by Kytka Hilmarová

All rights reserved. No part of this publication may be reproduced, distributed, or transmitted in any form or by any means, including photocopying, recording, or other electronic or mechanical methods, without the prior written permission of the publisher, except in the case of brief quotations embodied in critical reviews and certain other noncommercial uses permitted by copyright law. For permission requests or information about special discounts or bulk purchases, please contact:

Czech Revival Publishing.
www.czechrevival.com

The translated content in this work represents the views and opinions of the original author and is provided solely for educational and historical purposes. It is important to clarify that the translator and publisher do not endorse or support any form of anti-Semitic or anti-Freemason sentiment expressed in the text. We strongly condemn any form of discrimination or hatred towards any individual or group. Readers are encouraged to approach the content critically, engaging in thoughtful analysis and forming their own conclusions. It is crucial to promote understanding, respect, and tolerance in all discussions and to reject any form of prejudice. This work aims to contribute to a broader understanding of historical perspectives without promoting hatred or discrimination against any particular group, including the Jewish people and Freemasons.

Library of Congress Cataloging-in-Publication Data

Jacobi, Walter 1919 – 1947
Hilmarova, Kytka 1964-
    Golem - The Broom of the Czechs: The Decay of Czech Nationalism/ Walter Jacobi, translated by Kytka Hilmarová.

Summary: "Golem: The Broom of the Czechs" is a controversial book written by Walter Jacobi, a Nazi war criminal. The book promotes anti-Semitic and anti-Freemason ideologies, alleging that Jews and Freemasons were responsible for undermining Czech nationalism. It explores these ideas through the lens of the Prague Golem legend and examines historical events and individuals in relation to Czech identity. The book has garnered attention for its inflammatory and discriminatory views, but it provides insights into the political climate and propaganda of the time. – Provided by publisher.

ISBN 13: 978-1-943103-18-8

1. History / Europe / Eastern 2. Social Science / Freemasonry & Secret Societies 3. Political Science / Political Ideologies / Nationalism

"Anti-Semitism is a noxious weed that should be cut out. It has no place in civilization."

— *William Bruckner, Holocaust survivor*

This book is dedicated to the memory of Jan Opletal (December 31, 1914 - November 11, 1939), a courageous student at the Medical Faculty of Charles University in Prague. Jan Opletal's life was tragically cut short, but his unwavering spirit and determination in the face of Nazi oppression have made him an enduring symbol of Czech resistance. On October 28, 1939, during an anti-Nazi demonstration held on Czechoslovak Independence Day, Jan Opletal was shot, sustaining severe injuries that ultimately led to his untimely death two weeks later. His sacrifice and bravery continue to inspire generations, serving as a reminder of the indomitable spirit of those who fought against tyranny.

Contents

Preface ix

Introduction 1

Domestic Resistance 9

The Masonic Veil of Maffie 31

Ensuring Secrecy 48

Rosicrucians and Knights Kadosh 63

"The Light" in the Sokol Organization 81

The Temple of Youth 105

Semitic Czechoslovaks 123

Golem above the Castle 143

Illustrations from the Book 165

About the Author 221

About the Translator 227

Translator's Endnotes 229

Preface

Anti-Semitism is not just a belief or an opinion, but a grave affliction that poisons hearts and minds. It perpetuates hatred, sows division, and destroys the very fabric of our society. Yet, we hold within us the power to combat this darkness, to reject prejudice, and to forge a world where every individual is cherished and respected for their inherent worth. Let us unite, side by side, and proclaim with unwavering resolve: Never again. Never again shall we allow the forces of hatred to prevail. Never again shall we remain silent when faced with injustice. Never again shall we allow the shadows of bigotry to overshadow the radiance of empathy. Together, we possess the capacity to shape a future where love triumphs over hate and enlightenment conquers ignorance.

I had the privilege of receiving a generous gift: two sizable boxes filled with various Czech books, one of which was this 132-page hardcover edition that caught my attention because of the word Golem on the cover.

As I held the book in my hands, its slim and delicate appearance initially suggested a children's story centered around the mysterious Prague Golem. Yet, my interest was instantly stirred by the emblem adorning the cover – the square and compass, a symbol widely associated with Freemasonry, although this particular depiction seemed unfamiliar. (Later research would unveil its disguised portrayal as the Star of David.) Coupled with the captivating subtitle "The Decline of Czech Nationalism," I felt an undeniable pull to delve into its pages, immersing myself in a continuous reading session, driven by a yearning to uncover the intricacies of Czechoslovakia's historical narrative.

It quickly became apparent that this work held considerable historical value, offering unique insights into a specific period and topic.

After reading it, I immediatly knew it was crucial to approach the material with a critical and discerning lens, recognizing the author's obvious underlying agenda. The fact that the book was authored by a German Nazi during a period of intense anti-Semitic ideology raised significant concerns and calls for careful interpretation.

In November 1941, Walter Jacobi, the head of the SD in Prague, authored a highly controversial anti-Masonic publication titled "Golem...Geissel der Tschechen, die Zersetzung des Tschechischen Nationalismus" (Golem – The Scourge of the Czechs, the Disintegration of Czech Nationalism). This work was commissioned by Reinhard Heydrich and published by Orbis Verlag in December 1941. The book gained significant popularity, going through multiple editions and being available in both German and Czech languages, though it had never been translated to English.

I have to note that understanding the historical context is imperative in comprehending the motivations and biases of the author. It is evident that the author's perspective was deeply influenced by the pervasive anti-Semitic sentiments prevalent in Nazi Germany during that time. As such, it is essential to acknowledge the potential distortion of facts and the deliberate propagation of harmful stereotypes that may be present within the book's pages.

By acknowledging the author's affiliation with the Nazi regime, we can navigate the material with a heightened awareness of the inherent biases and prejudices. This critical approach allows us to engage with the text in a manner that ensures a more nuanced understanding of the historical events described, while also recognizing the dangerous ideologies that underpin its content.

In addition, when studying such works, it is crucial to confront the dark chapters of history and examine the motivations and perspectives of those involved, even when they are deeply troubling and contrary to our values. By doing so, we equip ourselves with the necessary knowledge to challenge and refute such harmful ideologies, ensuring that history does not repeat itself.

Therefore, it is important to approach the content of this book critically, given the author's association with the Nazi security service. However, it is worth noting that despite this affiliation, the passage presents a surprising degree of objectivity in its description of events relating to the Czech nation. The author argues that Czech nationalism, influenced by factors such as Zionism, Freemasonry, and philanthropic humanism, as well as the collective outcome of liberal democracy, was ultimately destined to falter shortly after its political emergence.

The book further distinguishes between nationalism, which encompasses a deep love and attachment to one's nation, its culture, and traditions, and patriotism, defined as loyalty to the state and its regime. It suggests that proponents of Masaryk, Beneš, Karel Čapek, international Jewry, and world humanism valued patriotism over nationalism. This perspective raises questions about the dynamics between these different ideologies and their impact on the course of Czech nationalism.

As a Czech historian and translator of books, I decided to translate this book because I believe that every aspect of history, even the troubling ones, deserves to be examined and understood. While I acknowledge that the book contains anti-Semitic content, it also offers insights into the prevailing thoughts and ideas of that time. I firmly believe that history should not be buried or ignored; instead, by confronting it directly, we can gain a deeper understanding of the motivations and beliefs that led to such atrocities.

As an individual, I want to make it clear that I do not endorse or support any discriminatory or prejudiced views expressed within the translated text. While engaging with this analysis, I believe it is important to approach it with a critical mindset, taking into account the potential bias inherent in the author's viewpoint and the historical context in which it was written. (Remember, Walter Jacobi was a SS officer holding the rank of SS-Obersturmbannführer, and became notorious as the head of the Sicherheitsdienst (SD)[1] in Prague.)

---

[1] The SD refers to the Sicherheitsdienst (Security Service) which was the intelligence agency of the SS (Schutzstaffel) in Nazi Germany. The SD played a crucial role in gathering intelligence, conducting surveillance, and carrying out security and political activities both within Germany and in occupied territories during World War II. It was responsible for monitoring and suppressing perceived threats to the Nazi regime, including political opposition, espionage, and dissent. The SD also played a major role in implementing the Holocaust by coordinating and executing the systematic persecution and extermination of millions of Jews and other targeted groups.

Introduction

This translation is based on the book titled "Golem: metla Čechů...: rozklad českého nacionalismu," which was originally published in Czech in 1942. The author's original essay was titled "Golem...Geissel der Tschechen, die Zersetzung des Tschechischen Nationalismus" and was written in German. While different translations of the title exist, such as "The Plague of the Czechs" or "The Scourge of the Czechs," the chosen translation of "The Broom of the Czechs" metaphorically represents the act of sweeping or cleaning. In this context, it symbolizes the critical examination of Czech nationalism and its decline.

Thye book delves into a thorough examination of the author's analysis regarding the various historical and social factors that contributed to the weakening and decline of Czech nationalism. It provides valuable insights into the intricate dynamics and challenges faced by the Czech nation as it grappled with its political landscape and the quest for a cohesive national identity.

It is crucial to bear in mind the background and context surrounding the book. The author, a German lawyer and SS officer named Walter Jacobi, was directly involved in the Nazi regime, and committed war crimes.

He was commissioned by Reinhard Heydrich[2], a high-ranking Nazi official, to write the book while serving in Prague. Jacobi's active participation in the Nazi Party, including his

---

[2] Reinhard Heydrich, a high-ranking SS officer and one of the key figures of the Nazi regime, played a significant role in implementing and overseeing the Holocaust during World War II. Known for his ruthless methods and unwavering loyalty to Hitler, Heydrich was appointed as the head of the Reich Main Security Office and later as the Deputy Protector of Bohemia and Moravia. His brutal rule and oppressive policies in occupied territories earned him the nickname "The Butcher of Prague." Heydrich's assassination in 1942 by Czech resistance fighters led to severe reprisals and further intensified Nazi brutality.

involvement in the SA unit and his role in executing individuals who opposed the regime, highlights his deep involvement in the atrocities committed during that time. His association with the massacre of Lidice further underscores his active participation in heinous acts. These facts shed light on the potential biases and distorted perspectives that may be present in his writings.

Be aware that Jacobi makes disparaging and incendiary claims, attributing malicious motives to Jews and assigning them blame for undermining Czech culture and national identity through enforced assimilation and multiculturalism. It asserts that these actions are part of a broader conspiracy orchestrated by Jewish immigrants from the former Soviet Union, who allegedly employ deceptive rhetoric of equality, freedom, and brotherhood to advance the dismantling of nations, the erasure of racial and cultural distinctions, and the establishment of a worldwide Jewish state.

Jacobi employed skillful tactics to convey his message, suggesting that the traditional Czech nation and its prominent movements like Sokol or Junák (Scouts) were originally virtuous. However, he argued that these entities had been co-opted by Masons, who were themselves infiltrated by Jews.

During the late 1930s, as Germany sought to undermine Czechoslovakia, Nazi propaganda aimed to discredit the Czechoslovak state before its eventual occupation and annexation. Despite its flaws, Czechoslovakia remained the last bastion of liberal democracy and the rule of law in Central Europe. Prominent individuals of that time were ardent patriots who wholeheartedly embraced the principles of their modern republic. Many found solace in the moral teachings of Freemasonry, which further fortified their social and political beliefs. It is important to note that the influence stemmed from their own convictions, rather than Freemasonry dictating their actions as alleged by the Nazis.

According to Jacobi, these Masons served the interests of cosmopolitan and plutocratic powers, acting as an international Golem. He contended that the election of Freemason Beneš as president in 1935, accompanied by the presence of Masons in key positions within the state, army, academia, Sokol, Junák, and cultural institutions, allowed the Golem to gain complete control over the Czech people, leading to the destruction of their national spirit. Jacobi claimed that removing Beneš and his alleged "corrupt Masonic clique," who he believed served world Jewry, would free the Czech nation from this harmful influence, allowing it to regain national esteem under German protection.

However, the actual situation was far from Jacobi's portrayal. The number of Jewish members in Czech lodges was minuscule, and Czechoslovakia remained the sole democracy in Central Europe during the late 1930s, where democratic leaders were drawn to Freemasonry, which continued to operate. Contrary to Jacobi's insinuations, lodges did not appoint their members to public positions.

The Czechoslovak Freemasons who held influential positions in the state were unable to fully achieve their aspirations for a free society guided by the moral values of Freemasonry. Their efforts were hindered by external factors and historical events that were beyond their control. Despite their genuine commitment to promoting positive change, the circumstances of the time prevented them from fully realizing their vision for a society built upon the principles they held dear.

The book also explores the legend of the Golem from Prague, a mythical creature created by the Mahara'l in the 16th century to protect Jewish communities from persecution, as an example. However, in Jacobi's context, the Golem was created with the intention of destroying Christians and other adversaries of the Jews, thereby establishing Jewish dominion over the world.

Jacobi alleges that Jews, acting as the Jewish "Golem," are internally opposed to the nationality in which they reside, specifically targeting the Czech people. He argues, "The old Jewish legend about the Golem, 'the real Prague man,' took on a new meaning during the twenty years of Czech sovereignty. The Golem that enslaved the Czech people defeated them through spiritual confusion, rendering them unable to liberate themselves from Jewish oppression."

The original book featured a cover depicting a statue of the Mahara'l created by Ladislav Jan Šaloun in the Prague City Hall, displayed since 1917. Additionally, it includes photographs of the Jewish cemetery in Prague and various documents intended to incriminate Jews.

As a passionate student of Czech history, I find this book particularly significant as it uncovers lesser-known facts and events that played a pivotal role in the decline of Czechoslovak nationalism during the First Republic. By delving into these lesser-explored aspects, the book provides valuable insights into the complexities and forces at play during that crucial period of Czech history.

The book explores a wide range of subjects, including the speeches of Svojsík, the role of Freemasonry in Czech nationalism, significant historical events surrounding Sokol and the Olympics, and the notable contributions of figures like Jan Masaryk, Dr. Jan Kapras, and Alphonse Mucha. By delving into these diverse topics, the book provides invaluable insights into their historical significance and deepens our understanding of Czech history. It offers a nuanced perspective on the factors that contributed to the decline of Czechoslovak nationalism.

Upon its release, the book was subjected to mandatory reviews by all Czech newspapers under Nazi control, who were instructed to hail it as a literary triumph.

Now, for the very first time in English, I present the translated version of this work.

Domestic Resistance

In 1928, a small booklet titled "Z domácího odboje" (From the Domestic Resistance) was published in Prague. It had barely thirty pages. The publisher was credited as "Památník odboje" (Memorial of Resistance) in Prague, and it was printed by the printing press of the former Ministry of National Defense in Prague-Bubeneč. Later that same year, the pamphlet was translated into French with an expanded title, "La Lutte Intérieure pour l'Indépendance Tchécoslovaque" (The Internal Struggle for Czechoslovak Independence). As the title suggests, this treatise provides a brief overview of the political behavior of Czechs at home during the four years of the World War.

The author of the booklet is none other than the long-time Prague law professor, Dr. Jan Kapras. His position as a scholar and national politician guarantees, at least in terms of describing actual events, that he approached his reflections critically and judiciously and was especially aware of his responsibility when making politically significant statements about the aforementioned period. The ten-year time gap since the events described allowed him to assess matters with less bias, and it also worked in his favor that important and mostly extensive works of Czech revolutionary literature had already been written, making it possible to extensively review the relevant material. Therefore, we can accept his account as all the more significant.

Kapras begins his treatise by stating that until the war, there were no preparations for a revolutionary solution to the Czech question, and that the Czech nation had neither a long-standing major political line nor a unified direction in terms of statehood. However, when the World War broke out, the anti-war sentiment among the Czechs was clear. Although "it was clear to everyone from the beginning that we couldn't conduct a revolutionary war at home, so there was no significant resistance during mobilization, occasional resistance arose later at home, which was punished with several executions. The focus of domestic politics at that time was based on the slogan 'not to fight and surrender,' and later on the expectation of the Russians" (Kapras, p. 4).

With this, the author already indicates one main line of Czech behavior during the war: political passivity towards the state, the government, and the needs of the war, which dominated the people and continued to dominate them for the next four years. As the war dragged on, it became more necessary to counter and thwart the attempts of some activist politicians to win the Czech people over for cooperation and a positive stance towards the state. "For this passivity, broad layers of the Czech nation and the majority of outstanding politicians, led by Kramář and Rašín, had an instinctive inclination" (p. 5).

It was therefore understandable that the Austrian General Staff, based in Těšín, completely different from the sluggish Viennese government, quickly became biased against the Czech regions of the monarchy and was further strengthened in its political doubts when, around the New Year of 1914/15, Czech units showed their unreliability for the first time, and when the entire 28th Infantry Regiment surrendered on the Russian front. The necessary intervention of state authorities was more pronounced in Moravia and Silesia than in Bohemia, where the Governor František Count Thun, based in Prague, almost reinforced Czech behavior by considering it harmless. Thun later had to yield to the pressure from politically more astute military circles and was replaced by Count Max Coudenhove in April 1915. However, Coudenhove, lacking initiative, developed into a mere bureaucratic executor of Vienna's orders. The government's attempts to win over Czech politicians – the possibilities of which were favorable, especially among the clerics and social democrats – as well as the call in the press to finally abandon the politics of passivity and for the Czech nation to declare itself clearly and uncompromisingly in favor of the Habsburg dynasty and the state, had no result thus far.

This fundamentally passive behavior of the majority of Czechs seemed to require secret guidance and leadership. While Kapras does not explicitly emphasize such a causal connection, he describes how secret organizations began to form precisely at that time, which logically followed the unified Czech sentiment and the government's attempts to undermine it. He explicitly states that several "secret circles" had already formed a few months after the outbreak of the war, around which Czech politics was centered. Among them were groups of proponents of statehood who sent L. Sychravý as a liaison to Switzerland, and realists, with Masaryk as their leader, who went to allied countries with a

Western orientation and saw possibilities of politically decisive help in France and the Anglo-Saxon countries. Meanwhile, the Young Czechs, led by Kramář and Rašín with strong influences of such thinking in the background, looked to the East and leaned towards hope in Russia.

Among the leading representatives of these groups, a common secret working committee called the "Maffie" was soon formed[i]. Initially, it included Kramář, Rašín, Hajn, Sámal, Soukup, Franta, Scheiner, and Beneš. Later, when some members fled (Beneš) or were arrested (Kramář, Scheiner, Rašín), many others joined the committee.

In describing the activities of the Maffie, whose central headquarters were precisely in Prague, Masaryk referred to it as a necessity. Kapras limits his account to a brief sentence: "Maffie maintained constant contact with the foreign resistance, especially with Masaryk, initially through Beneš, and when he also went abroad, through numerous individuals whose records were concentrated in the hands of the Bělehrádek group" (p. 6/7). With this, the Czech intelligentsia took the lead in rallying the popular masses, who were inspired by passivity and the will to resist. "Our domestic resistance, in its beginnings, was a resistance of broad unorganized popular masses who suddenly realized that a great moment had come for the Czech nation. It was the resistance of thousands and thousands of ordinary and unknown people, many of whom endured suffering for their convictions... Our domestic resistance is therefore primarily the resistance of the people who heroically endured the burdens and sufferings of war, hoping for a better future. Secondly, it is the resistance of the intelligentsia that understands its place is at the forefront of the people" (p. 26).

The resistance activities of the secret Czech leadership, now directed and systematically organized from the top, took two directions. Firstly, through political sabotage, they had to take a stand against the government in Vienna and thwart any attempts by official policies to undermine their cause. Secondly, the activities of the Maffie turned against the plans and efforts of Czech activists who sought to independently influence the fundamental position of the Czech nation. In this context, Kapras presents the following remarkable sentence as the guiding idea behind this disruptive activity: "Under these circumstances, it seemed that the task of domestic politicians was merely to remain silent and prevent domestic actions from spoiling and interfering with actions abroad." (p. 8).

When attempts were made at the end of 1915 to unite the various parties, and in January 1916, under the leadership of Dr. Matěj Dra, a far-reaching merger of the Old Czechs, Young Czechs, Realists, and National Socialists into a unified Czech party called the "National Party" began, this party was immediately sabotaged by the leadership of the resistance. There were concerns that "the government would impose its demands, especially those aimed at Czech foreign policy, more easily on this unified party than on the previous several parties" (p. 8). This danger was so ominous that it was precisely at that time that the Vienna government actively acted against the effects of Czech emigration activities at home and suppressed the influence of Czech propaganda from abroad. Caught between the desires of the government and the undermining work of resistance groups, the "National Party" collapsed in May 1916.

However, the danger of a more friendly Czech policy towards the state was not definitively extinguished. Several months after the dissolution of the unified Czech party, a new unified political organization took shape in the form of the "Czech Union," which was formed with decisive cooperation from Švehla of the Agrarian Party, which now had a stronger presence in the political arena. All parties, except for the statehood-progressive and realist ones, were consolidated here, including their parliamentary clubs.

The underground struggle against the activists resumed, especially when the "Czech Union," following the diplomatic act of the Entente on January 10, 1917, which mentioned Czechoslovak liberation, and President Wilson's declaration on January 22, 1917, regarding the right to self-determination of nations, submitted a letter to the Austrian Minister of Foreign Affairs that amounted to a complete renunciation of Czech foreign activities. Kapras clearly condemns this "complacency of the union, which caused a sudden decline in its credibility in Bohemia" (p. 11-12), and he describes how futile the union's efforts were to save its shaken position in Vienna by demanding a return to parliamentary government forms. "It is understandable that in this state of affairs, there were very serious concerns about any expressions by Czech deputies and their Czech Union representatives, and the goal was to prevent these expressions" (p. 12).

For this purpose, as Kapras explicitly emphasizes, the "Národ" group was formed at the beginning of 1917, consisting of like-minded individuals from the ranks of the Young Czech party opposition, statehood proponents, realists, or politically uninvolved figures in Czech cultural life. "They were all determined not to allow Czech politics to miss the irrevocable opportunity to free the Czech state from Habsburg subjugation. The publication of the first issue of the 'Národ' group's organ on April 8, 1917, was a significant event in the prevailing circumstances, as it openly and consistently proclaimed strong resistance against the previous policies of the Czech Union" (p. 13).

Until 1918, the "Národ" group played a significant role in political events. (Kapras deliberately conceals his membership in this group, just like his other active participation in the domestic resistance.) Subversive activities also began soon after, when, for the first time during the World War, the Viennese Imperial

Council was convened on May 30, 1917, and there were concerns that Czech deputies might not act appropriately in this historic moment.

On May 17th, that famous speech by Czech cultural workers was published, written by Jaroslav Kvapil, the head of the drama department at the Czech National Theater. It was signed by 222 representatives of literature and science as a statement to the "Czech Delegation in the Imperial Council." The speech called on Czech deputies to develop and defend the Czech program in front of the entire European community. Kapras celebrates this action with enthusiastic words as the first major political expression of the Czech domestic resistance, which was joyfully embraced by the entire Czech nation and correctly understood by the people as "a demand that had long been advocated by Czech propaganda abroad" (p. 14).

Then, party political life began to be revived. Within the Young Czechs, an uncompromising opposition direction emerged, centered around Franta, a member of the Maffie, and the aforementioned František Sísa, who was sidelined after Kramář and Rašín were arrested. The same occurred within the National Socialist

Party. Among the Agrarians, the radical wing gained prominence, which was not surprising because it was precisely here that the Maffie's tactic of establishing early political bonds through knowledge of illegal activities proved successful. Kapras mentions on the first pages of his work that Kramář and Švehla sent Deputy Diirich illegally to Russia in 1915.

However, it is only in Beneš's book, " Světová válka a naše revoluce" (World War and Our Revolution), that the final explanation is provided when describing the Maffie meeting held in the spring of 1915 after Beneš's return from his trip abroad to meet Masaryk. Beneš states, "Everyone agreed on the mission of Dürich, also because he was a deputy of the Agrarian Party, and that it would be discussed with Chairman Švehla, thereby formally incorporating this powerful party into the conspiracy" (Beneš, p. 66). Thus, the course of the major party of Czech farmers was set for the duration of the World War, and if necessary, this party could be compelled to join the ranks of the domestic resistance's underground front.

All of these circumstances together resulted in significant political pressure on the Czech representatives until the opening of the

Viennese Imperial Council on April 30, 1917, when they mustered enough strength to demand territorial autonomy in line with Czech national law. The official declaration of the Czech Union was received with relief among the Czechs because the people had a keen sense for the underlying statehood formulations of their leadership. The representatives of the statehood-progressive party issued their own statement, which did not mention Austria at all. "This uncompromising declaration was exactly what the entire Czech nation wanted" (p. 16).

The Czech representatives then adopted a new tactic for their political approach. When a commission for the reform of the pre-Litavka constitution was established in Vienna, they decided not to bypass its meetings and avoid political responsibility. To justify this behavior externally, a compromise proposal from the Agrarian Party was accepted, stating that they would not participate in the sessions of the constitutional commission until discussions on the principles contained in the statehood declaration of May 30 were concluded. Several weeks later, the majority of party representatives pushed the situation to the brink with an even more far-reaching resolution to officially protest against any participation in the

Vienna discussions. Under Klofáč's leadership, the Statehood Club was formed, with the aim of "preventing opportunistic deputies, especially the Young Czechs, from advancing their agenda" (p. 18).

The increasingly visible shift in the internal power dynamics within Czech politics in favor of the uncompromising faction led to "Národní listy" falling into the hands of one of the most radical representatives, František Sísa, and soon also to Rašín, a leading member of the Maffie, upon his release from prison, providing them with new opportunities for influence. When Kramář was subsequently released in October 1917, his release was seen as a clear sign of the state's weakness, and this politician was triumphantly welcomed as the main representative of the domestic resistance. Reports from abroad increasingly became a guiding force. "All of this rapidly strengthened the anti-Austrian radical direction in the autumn of 1917, naturally reinforced by news from abroad" (p. 18).

In broad strokes, the merging of the two groups that would later form the National Democrats, led by Kramář, and the National Socialists, led by Klofáč, was already taking shape. Even within the Social Democrats, a radical faction led by Habrman was gaining influence.

At the beginning of 1918, events gained momentum. On Three Kings' Day, January 6, 1918, the General Assembly of Czech Deputies, which was no longer restrained, demanded the establishment of a separate Czech state and made no mention of the Austrian state union. On April 13, a solemn oath ceremony of all layers of the Czech nation took place in Smetana Hall in Prague, where Jirásek, amidst great attendance, read the wording of the oath to the independent Czechoslovak nation. In May, large theatrical celebrations were organized in Prague, attended by representatives of Slovaks, South Slavs, as well as Polish and Italian minorities. A festive meeting was then held in the Municipal House in Prague. "The May demonstrations were viewed in Vienna as an organically connected part of the Czech foreign resistance and triggered the final outbreak of political persecution" (p. 21).

On July 13, 1918, after preparations, a new National Committee led by Švehla, the leader of the Agrarian Party, was formed. Its task from the beginning was to declare the independence of the Czech state at the opportune moment and make all necessary preparations for it. Finally, on October 2, the Czech mission in Vienna openly declared its allegiance to the resistance. Later, on October 24 and 25, the Czech delegates traveled to Geneva to meet with Beneš. On the night of October 27-28, Dr. Scheiner, the mayor of the Sokol organization, who was also released from prison, organized guard duty. The next day, a group of four, under Rašín's leadership, convened and decided to declare the independent Czechoslovak Republic upon Austria's surrender. That same evening, the first law of the new state was issued. The Sokol organization took charge of order and security services.

The assumption of the first executive power by Sokol represented the culmination of their tireless underground work during the World War. However, Kapras's document mentions their activities only briefly. Sokol is mentioned only when listing secret resistance groups, and the activities of its leader, Dr. Scheiner, in Maffie are mentioned several times.

However, the missing details can easily be supplemented with information from the memoirs of Masaryk and Beneš, as well as from

German publications by Vienna State Librarian Dr. Pavel Molisch. According to these sources, Masaryk established contact with Dr. Scheiner, the leader of Sokol, immediately after the outbreak of the war. Masaryk mentions in his book "Světová revoluce" (The World Revolution), that they agreed that Sokol members would serve as a national security guard during the Russian occupation, akin to a national army, as needed (Masaryk, p. 14).

At the same time, negotiations were held regarding the financing of Czech activities abroad using Sokol funds, and Dr. Scheiner immediately provided the first contribution to initiate their work. This was followed by the state's first interventions against individual gymnastic units, including Sokol in Jičín. When Masaryk departed for abroad, financed by Sokol funds, he entrusted his collaborator Beneš, who remained at home, with his secret contacts with Sokol. Masaryk instructed Beneš to establish contact with Dr. Scheiner and ensure that all further activities were coordinated with him. Masaryk had great confidence in Sokol and expected that funds for revolutionary activities could also be obtained from there. He also believed that in preparation for a potential violent revolution, it was necessary to prepare Sokol leaders and the Sokol masses. At that time, Dr. Scheiner himself contemplated fleeing abroad and wanted to go to America to stimulate action among compatriots. He pretended to purchase cotton for the Bohemia Bank in America and requested a passport from the police. Despite repeated interventions, the police rejected his request and prevented him from leaving (Beneš, p. 27).

During the establishment of Maffie, Dr. Scheiner belonged to the initial circle and, through Sokol units, gathered a wealth of military information. He was later instructed to find a way to release Sokol funds, which had been immobilized due to the fear of confiscation, and their release was subject to the approval of individuals who could not be informed about their purpose (Beneš, p. 41).

Meanwhile, a series of events occurred in the Austrian army that seemed to demonstrate a certain planning behind the Czech soldiers' desertion among the defectors on the Russian front. From letters of war prisoners, it could be inferred that Sokol encouraged its members called up for military service to join the "Czech Comradeship," and it became apparent that soldiers of the Austro-Hungarian army defecting on the fronts carried Sokol identification cards, which particularly in Russia served as evidence of their anti-German sentiments.

Consequently, purely Sokol units were formed within the ranks of the Czech legionary regiments, which were formed on the side of the Allies. Masaryk therefore proposed that Scheiner escape to Russia and take charge of organizing the Czech army there, but instead, Diirich illegally departed in his place. Since 1915, it had become increasingly clear that Sokol acted according to the instructions of the Paris Czech Central Committee. In May of the same year, Scheiner was arrested and accused by the Supreme Army Command. However, he skillfully played a double game, and due to lack of evidence, the proceedings were halted, especially since the authorities were unaware of his involvement in Maffie. Upon his release, he immediately returned to the inner circle of Maffie, and when the United States declared war, he once again provided reports to Dr. Beneš abroad. Thus, his example led Sokol's members to engage in continuous underground work, despite the strict surveillance by the authorities.

However, it was mainly the members of the older generations who remained at home and were available for resistance activities within Sokol. The youth were largely prohibited from participating in Sokol activities, and Czech schoolchildren were completely excluded from Sokol exercises by a state ban. Kapras does not mention the anti-state sentiments of Czech youth in his work, but it can be evidenced by a significant document from the revolutionary period. A. B. Svojsík, the first leader of the Czech Scout movement, presented an approved annual report to the General Assembly of the Junáci (Scouts) on November 28, 1918. The speaker gratefully recalled the understanding and support given by Masaryk, Kramář, Jirásek, Klofáč, and others to the idea of Junák, intentionally emphasizing his efforts to preserve the Scouts as a tool for Czech youth education only. "And at the moment when we were already preparing to put our plans into action and establish Czech scouting based on the American and English models, a proposal was submitted to the Czech authorities - let it be accused that it was submitted by a Czech person - to introduce scouting based on the models of German organizations, Wehrkraft and Pfadfinder, with all the distasteful military maneuvers and the notorious methods of educating well-Austrian-minded youth, directly recommended by officers." However, Kramář repeatedly managed to protect the Scout movement and find ways to avert the looming danger. "When these traps failed, they came up with another one. They established the Imperial Union in Vienna."

They offered complete autonomy, language equality, generous subsidies by our standards, positions in committees, even vice presidencies... Politely, but certainly, I refused, just like those who later arrived from Vienna in general uniforms or with baronial and count titles."

It was clear as day that with such sentiments among the leaders of Czech youth, no loyalty to the state could be expected from their units during the critical war years. This led to extensive acts of sabotage by the Junáci (Scouts), who repeatedly deceived the Austrian authorities and managed to avert dissolution. Outwardly, they engaged in humanitarian activities, organizing actions for wounded Czech veterans, supporting orphans and widows, assisting needy writers, artists, and students. However, their involvement in state-sponsored war support initiatives was rejected.

Svojsík describes this situation with telling words: "Austria wanted to replenish its resources with metal, rubber, rags, and other scraps, but we did not allow a single scout to participate in these war-supporting activities. Despite the fact that not only the youth of primary and secondary schools but also high school professors were directly commanded to contribute. We also avoided distasteful military exercises, and our scouts never heard a single German command." The speaker then points out how openly the "Junák" magazine addressed this issue. Educating children in the Czech Scout movement was not easy when it came to teaching them to fulfill obligations to the state and to circumvent laws and regulations. However, "the significance of our nation's silence and behavior during the war was clear, especially to the children, and even more so to the scouts. Every leader found an opportunity to speak clearly and show that artificially imposed obligations to our so-called homeland stood in opposition to the holiest duty: duty to the Nation, Democracy, Freedom. And for every Czech individual, there was only one possible choice."

By engaging in this activity, which I would call treason against the establishment, we managed to maintain the operation of our units within the limits compatible with our own convictions. We succeeded in preserving scouting, working diligently, and accumulating experience for the time when we would be able to organize scouting for ourselves." Based on this wartime activity in the Czech youth resistance, Svojsík therefore justified at the end of 1918 that the Junák movement should become an educational force for Czech youth alongside the Sokol.

The Masonic Veil of Maffie

In the first twenty years of their own statehood, the Czechs became accustomed to considering the era of domestic resistance during the World War as one of the most glorious chapters in their history. This period and its remarkable personalities stood in the collective consciousness of the broad populace as a monument, and in schools, year after year, they were portrayed to the new generation as a heroic era. Yet, in doing so, everyone contented themselves with a superficial view of the external events and anecdotes, losing sight of the critical analysis and the effort to uncover the underlying matters. These likely remained hidden even from many journalists and writers in Czech literature about the World War and the revolutionary literature. Some were aware of them, including the author of the treatise "Z domácího odboje" (From Domestic Resistance), Jan Kapras. However, even he unfolded a curtain of grand events and the façade of well-known occurrences and personalities to conceal the secrets he knew were behind them.

At one point in his treatise, Kapras, however, inadvertently reveals that he knows everything in detail. Symbolically, he mentions the names Sis and Dadone in immediate succession. Around these two individuals, the historical backdrop and personal connections intertwine, which have been concealed from the Czech public by all who knew about them. The personality of František Sis, even today, is known to the Czech people only as a figure of a deceased compatriot who, as Kapras also mentions, played a not insignificant role in the domestic resistance during the World War and was a Prague journalist for decades.

He was born in 1878 in Maršov, Moravia, and as a young student, he belonged to the most radical Young Czechs. In 1897, after the repeal of the language ordinances of Badeni, he organized anti-German demonstrations by Czech university students in Prague, and for these merits, he was elected to a committee the following year, tasked with securing permission for the establishment of the second Czech university. Alongside his active political career, he soon began working as a journalist. In 1907, he was appointed the General Secretary of the Young Czech (then National Liberal) Party, where he became the closest, albeit often headstrong, collaborator of Karel Kramář. Through his brother Vladimir in Sofia, he established confidential contacts with Russian Foreign Minister Sazonov in the years before 1914. When the World War broke out, he was able to supply crucial information of both civilian and military nature to the French and Italian military authorities through foreign embassies in Vienna, which were then passed on through Bulgaria. It was also through him that the first Czech memorandum, seeking the disintegration of the Habsburg monarchy and the independence of the Czech state, was sent to the Allied powers in March 1915.

Sis was the main representative of the uncompromising faction within the Czech Maffie. In his stubbornness and audacity, unmatched among the spokespeople of the domestic resistance at the time, he openly criticized Czech activist politicians and their hesitant behavior. He had connections on all sides, and with their help, he managed to secure permission to publish the magazine "Národ" (Nation) after the suspension of the "Národní listy" (National Letters or National News) newspaper. After the revolution, he was appointed the editor-in-chief of the revived "Národní listy," became a member of the revolutionary National Assembly, and later served as a National Democratic deputy in the

Prague Chamber. In 1921, he went to Paris for two years, where he founded the "Comité Slave." When Rašín was assassinated, he was once again called to his former position as editor-in-chief, and his personal fate became intertwined with the declining National Democratic Party. From 1930, his health deteriorated, and he passed away in August 1938, shortly before the outbreak of the Sudeten crisis.

František Sis had become acquainted with a foreigner living in Prague named Ugo Dadone years before the outbreak of the World War. Dadone, who taught him Italian, immediately initiated political connections with Sis when the war broke out in 1914. Dadone revealed himself to be a fervent supporter of the Italian irredentist program within the Habsburg monarchy and stated in conversations that his country would soon join the war on the side of the Allies against the Central Powers. Sis quickly developed trust in Dadone and introduced him to Rašín and other members of the Maffie. An eyewitness reported: "Dadone was the archetype of a conspirator. Optimistic and sanguine, with a great deal of cunning and Machiavellianism. He always recommended to us the tactics of Italian conspirators against

Austria. He explained how the leaders of the Italian conspirators would swear their loyalty to Austrian General Bubna in Milan until midnight, then after midnight, they would attend a conspiratorial meeting to discuss how to destroy Austria and pledge their love and loyalty to Italy." Thus, a political agreement was quickly established: Dadone provided foreign newspapers and war reports to the Maffie, while the Maffie provided him with information about events in Bohemia.

Dadone cannot be considered a true Italian nationalist; rather, he should be seen as an unrooted international conspirator who used diplomatic courier services for the purposes of the Freemasons. He was one of the confidants of the Roman "Gran Loggia Nazionale" and was regarded as a mediator of Italian Freemasonry, which was eradicated by Mussolini after 1923. It can be argued that Sis, through his association with Dadone, became more intimately acquainted with the issues of Freemasonry and that these connections had a strong ideological influence on him. However, his plan to establish a Czech lodge was primarily driven by Czech motives and likely stemmed from his own activities. Sis based his ideas on thoughts that he later recorded. (*The information presented in this chapter and the following two is partially based on unpublished documents.)

During the feverish and tumultuous times at the end of 1914, in the midst of the heaviest oppression of the Czech nation, when military absolutism reigned supreme over the lives and souls of Czech citizens, the idea of establishing a Czech Masonic lodge in Prague was born. We began to organize underground revolutionary activities more firmly, forming a network of secret agents to paralyze and endanger the functioning of the state. We focused on gathering intelligence about military affairs, war industries, the economic and financial state of the monarchy, railways, and all matters related to the war. It was during this time that the idea came to me that a Masonic lodge would be the most effective organization, both ideologically and practically, for the goals we had set for ourselves - the liberation of the nation and cooperation with the Allied powers.

Ideologically, the basic doctrines and goals of Freemasonry pointed the way, with its grand principles of liberty, equality, and fraternity, and the participation of Freemasonry in the liberation revolutions of oppressed nations.

Freemasonry has always stood at the forefront of the struggle for national freedom and human rights in all nations. The idea of liberation and inspiration led us into the underground, where we formulated plans, programs, and actions for the liberation of our nation. We deeply believed that this war was a decisive moment that would determine the restoration of Czech independence, but that the nation itself must stand firm, resolute, and embark on the path of work and deeds towards the historic day when we would stand among the independent nations.

Sis soon shared his plans with his party colleague and comrade in the Maffie, Dr. Rašín. Rašín approved his proposal to establish a Czech lodge of Freemasons, as it would provide a legal platform for Maffie meetings, where under the guise of Masonic activities, the work of the Czech resistance could be carried out. Two aspects influenced Sis and Rašín, as well as the circle of people they initiated. Firstly, they regarded the "traditionally and almost legendary guaranteed secrecy of proceedings in Freemasonry, whose lodges and members had endured countless persecutions by state and religious authorities, papal curses, inquisitions, imprisonments, and capital punishments, creating a legendary atmosphere of cohesion and trust." Secondly, they hoped that the Czech revolutionary lodge would find opportunities to connect with Freemasonry in other nations and that they would be able to "bring the Czech question to the world Masonic forum."

By expressing their aspirations within the framework of international conspiracy and supranational secret alliances, the circles of the Maffie made it clear that they did not want their resistance activities to be depicted as a popularly based national revolution, but rather as an international conspiracy. This concept already implied a world-view commitment to the principles of the French Revolution, with all its seemingly humanitarian consequences, which the Czech nation experienced firsthand during the misguided paths of its state politics in the twenty years of democracy. The praised architects of their own Czech state emerged as spokespersons for a spiritual movement that, from the moment of its political inception, was willing to subordinate this idea to the declining international powers. From the perspective of the Czech people, this moment encapsulates perhaps the deepest tragedy of their great historic moment and the subsequent course of development, which brought them to a point that for too long seemed only gloriously promising.

Let's return to the depicted historical situation. In Austria at that time, unlike in Hungary, Freemasonry was prohibited. However, this prohibition was circumvented, and Freemasons gathered in humanitarian and charitable societies in places such as Karlovy Vary, Žatec, Liberec, Pilsen, and other locations. They performed their ritual lodge activities on Hungarian territory. For example, in Prague, there was an association called "Hiram," which belonged to the "Symbolic Grand Lodge of Hungary" and conducted its lodge activities in Bratislava.

This association was utraquist, meaning it brought together Czech and German members. Rašín obtained unfavorable information from Czech members of this lodge, whom he considered politically reliable, namely from Freemasons V. Červinka and Dra Třebický. The information stated that German Freemasons were susceptible to "the influence of war sentiment" and thus needed to be approached with caution. The informants also believed that not all Czech Freemasons from the lodge were suitable for Maffie's purposes and that some still belonged to the activist camp. It seemed doubtful, therefore, whether seven Masters Freemasons, who would be authorized to establish a new lodge, could be found among their ranks. Sis himself states: "Naturally, we could not count on Freemasons from the activist camp to establish a revolutionary lodge, so I considered whether there might be another way to access a Masonic lodge other than through members of the Prague lodge, as this path proved impassable."

So the idea arose to utilize the political connection with Ugo Dadone, who was known as a Freemason, in this direction as well. It was Sis who first asked him whether it would be possible to establish a revolutionary lodge in Prague with the help of Italian Freemasonry. Dadone wrote to the Grand National Lodge of Italy around the New Year of 1915, and he received a response stating that this plan could be implemented. Selected individuals could be accepted as members of the lodge in Rome and then establish a Czech lodge in Prague as "authentic Italian Freemasonry." Preliminary negotiations in Prague extended throughout January and February 1915 but were stalled due to developments in foreign politics. Dadone left

Prague in March, and in May, the war with Italy broke out.

At present, it is still unclear what thoughts the prospective lodge members entertained during the subsequent years of the World War. There is no documentation from that period. Perhaps the arrests of leading members of Maffie, which began in May 1915, significantly hindered their plans. It can be ascertained that no lodge was established, and the aforementioned plans were not revived until the collapse of the Habsburg monarchy. The resistance group around the newspaper "Národ" continued its meetings "in conspiratorial confidence" even after the October Revolution, as explicitly emphasized in one place, guided by the idea that the future fate of the nation was uncertain and that "secret formations that would remain guardians of national destiny" were necessary for all possible cases. In this atmosphere, a new breeding ground for Masonic plans was evolving.

It was therefore appropriate to return to the idea from 1914/1915 of establishing a lodge and build upon the earlier attempts. As a starting point, the group "Národ" set the goal: "To create a legal foundation of Masonic lodges for the protection of national freedom and the defense of human rights, to transfer the content and methods of underground warfare into the lodges, and to gather those who have proven themselves during the war and have been tested by the dangers of war."

These programmatic explanations were characteristic. Maffie's activities during the World War were understood as "proof" in the spirit of international Freemasonry. This aligns with Sis's recorded sentiments from the revolutionary era: "My program was for the lodge established from our war group to continue its work in the same spirit as the 'Národ' group during the war. Just as the war circle of 'Národ' and Maffie prepared the nation's freedom, the lodge would become a guardian of our young freedom in the liberated nation."

Freemasonry for the plan to establish a Czech lodge. Before his return to Prague, he independently discussed the details with the "National Grand Lodge of Italy" and the "Supreme Council of the Ancient and Accepted Scottish Rite for Italy" in Rome. He succeeded in persuading Grand Master and Sovereign Commander Raoul Palermi to support the

Czech cause. When Sis resumed his contact with Dadone in Prague, Dadone was able to inform him of the position of the Grand Lodge and negotiate the necessary terms.

The "National Grand Lodge of Italy" expressed its willingness to do "everything that is right according to the general statutes of Scottish Freemasonry" to establish a regular lodge in Prague, ensuring strict compliance with all formal provisions of the general statutes and avoiding any obstacles to recognition by other Masonic powers. On a written request from at least seven Czech applicants, the Grand Lodge would grant its approval through the authorization of the Supreme Council and allow the applicants to be "per procuram" accepted into the "Gran Loggia Nazionale." New members would then be issued their lodge certificates and enrolled with the Grand Lodge. In response to inquiries, the Grand Lodge even expressed its willingness to shorten the prescribed periods for attaining higher degrees and to grant higher ranks beyond the level of Master Mason in the Scottish Rite.

During the Christmas holidays of 1918, members of the Maffie involved in the war discussed plans to establish a revolutionary lodge in Prague. Present at the meeting were Dr. Alois Rašín, Dr. Přemysl Šámal, František Sis, Dr. Josef Scheiner, and Dr. Bohumil Němec. They unanimously agreed to join Freemasonry. Immediate efforts were made to recruit more individuals who shared their views, and discussions took place both at Sis's apartment at Na Smetance 6 and in the meeting room of the "Národní listy" newspaper. In addition to those who had already been appointed, the protocol mentions the following names: Ventura, Thon, Jindřich Čapek, Dvorský, Syllaba, Matys, Dr. Stašek, Nušl, Folprecht, Ing. Dvořáček, Emil Svoboda, Dyk, and Kapras. Other members of the "Národ" group were also invited. All those initiated into the plan expressed their approval and willingness to "seek the light."

In January 1919, the first meeting of selected members was convened in the meeting room of the "Národní listy" newspaper. Ugo Dadone was present, and František Sis chaired the meeting, as noted in Sis's records. The attendees included Dr. Němec, Ing. Dvořáček, Dr. Dvorský, F. Táborský, Dr. Syllaba, Jindřich Čapek, Josef Čapek, Karel Čapek, Editor Vilém Heinz, J. Ventura, Dr. Chotek, Dr. Matys, Dr. Fr. Stašek, Dr. E. Svoboda, V. Dyk, Dr. Hoch, Dr. Folprecht, Dr. Nušl, Dr. Borovička, Dr.

Pospíšil, Dr. Kapras, Dr. Babák, Dr. Vladimír Slavík, Dr. Thon, and Dr. Krofta.

When the meeting began, Dadone presented a written full power of attorney from the Supreme Council "Federazione massonica universale del rito scozzese e antico e accettato" and delivered an introductory lecture on the statutes and rituals of Scottish Freemasonry, the conditions for establishing a lodge, and the obligations that each seeker must fulfill. When asked if they were willing to join, all attendees responded affirmatively, submitted their applications, and took the oath of the first-degree Freemasons, administered by the Italian delegate. They also accepted the obligation of obedience to the Supreme Council of the Ancient and Accepted Scottish Rite in Rome. Later, Dr. Alois Rašín, Dr. Přemysl Šámal, and Dr. Scheiner also signed the documents. The report from Rome took a while to arrive, and it wasn't until March 1919, that it arrived with the following outcome:

1. All the individuals mentioned were "per procuram" admitted as apprentices into the "Gran Loggia Nazionale Or... di Roma, Valle del Tevere."

2. The prescribed "waiting periods for wage increases" (promotions) were shortened, and the appointed individuals were simultaneously granted the degree of Fellow Craft and Master.
3. The appointed individuals were authorized to establish a regular lodge of the Scottish Rite in Prague.

On March 21, written certificates were issued to all Master Freemasons (3rd degree) who had taken the oath and pledge. Eight days later, the election of officers and officials of the lodge took place. The following were elected:

Master of the Lodge: Dr. B. Němec
Deputy Master: Dr. L. Syllaba
First Overseer: Dr. E. Svoboda
Speaker: Dr. V. Dvorský
Secretary: Fr. Sis
Treasurer: V. Dyk
Ritual Expert: Jindřich Čapek
Tyler: J. Ventura
Ceremonial Officer: Dr. J. Thon
House Manager: [name not provided]

Thus, the activities of the Maffie concluded just five months after the establishment of the

Czechoslovak Republic, within the embrace of high-level Ancient and Accepted Scottish Rite Freemasonry, which now took over the work of the Czech resistance groups during the World War.

Dadone's words when handing over the written certificates were expressive, as he stated that all present were exempt from the symbolic tests prescribed in the initiation ritual. The Grand Lodge in Rome considered it "proven that the seekers demonstrated, through their lives and actions during times of danger in the struggle for national freedom, the character, qualities, and principles that are necessary for admission into Freemasonry." The newly appointed Master Freemasons responded with an appropriate analogy, recalling their activities within the domestic resistance group "Národ" (Nation), unanimously deciding to name the lodge they founded accordingly.

Ensuring Secrecy

It is not irrelevant to inquire why the circles from the Maffie, who were joining Freemasonry, specifically chose the Scottish Rite. After all, there was already a lodge of the Johannite Freemasonry, Lodge "Hiram," in Prague, which belonged to the "Symbolic Grand Lodge of Hungary." However, doubts arising from the war, regarding its Utraquist composition and politically activist mindset among its Czech members, had to be discarded after the revolution in 1918. Moreover, starting around the New Year of 1919, many Czech members began leaving the "Hiram" lodge with the aim of establishing a purely Czech Johannite lodge named "J. A. Comenius." This lodge was established on October 26, 1918, two days before the revolution. It immediately affiliated itself with the "Grand Orient of France."

Why the Czech resistance fighters from World War I did not join existing groups of Freemasons cannot be fully documented in all details today. According to the opinion of the members of Maffie, the Johannite Freemasonry saw its purpose "solely in practicing Masonic philanthropy." However, they were clear about their own goals, stating: "If our aim was solely to become Freemasons with humanitarian and philanthropic goals, as was the main, if not the only purpose of the pre-war Johannite Freemasons in Prague, we could have sought admission into the Johannite rite." However, their primary goal was not to establish a lodge for the cultivation of traditional philanthropic and humanitarian Masonic programs or for the dissemination of humanitarian ideals. The members of Maffie had much broader objectives from the beginning. "Our goal from the start reached much higher. It aimed at the pinnacle of Freemasonry, at the Masonic dogmas of Liberty, Equality, Fraternity," following the influence and activities of Maffie during World War I and aiming to preserve this tradition.

Essentially, according to the beliefs of the resistance fighters at the time, the ideas and teachings of both the Johannite lodges and Scottish Freemasonry were quite similar. They claimed that the customs and rituals were almost identical, with differences primarily concerning internal organization, which they considered decisive.
However, they emphasized the preference for Scottish Freemasonry, as it had a strong

presence and influence not only in Latin countries but also in Anglo-Saxon countries and America.

Nevertheless, Maffie rejected collaboration with German Freemasons, even beyond organizational matters. They fundamentally denied the possibility of seeing trustworthy partners among the Germans, and they believed that cooperation between Germans and Czechs was inconceivable, even on a Masonic basis, at least during the initial period. František Sis was a vocal advocate of such positions. In preparatory discussions among like-minded friends seeking to join the lodge, he explained that "even the same words have different meanings in Czech and German conceptions. The concept of honor is different, as is the concept of God. The entire spirit and tradition of our national history are in opposition to the spirit and traditions of German history as projected by the minds of all Germans."

In the eyes of Maffie, the Freemasons in Austria were no longer considered revolutionaries, and they had supposedly erased the memory of the 1848 year of freedom, in which they participated, from their recollections. Their action program allegedly no longer included the guiding principles of "Liberty, Equality, Fraternity," and they settled for "philanthropic humanism."

All the knowledgeable personalities of Maffie, and if applicable, the "Národ" group, unanimously decided, after several consultations, against the Utraquist lodge and against Johannite Freemasonry as a whole. They believed that the realization of their Masonic ideals was guaranteed only in Scottish Freemasonry (high-degree Freemasonry).

Finally, in the discussions, it was stated that only this solution was compatible with all potential political developments. It is interesting to note the impact of the contemporary events in foreign politics, such as the Bolshevik rule of Béla Kun in Hungary and the Marxist desire for dictatorship in Austria, on these consultations. There were voices of pessimism regarding the overall political situation and the future prospects of Czechoslovakia, particularly expressed by Dr. J. Matys, who was highly skeptical.

In order to exclude Germans at all costs, Emil Svoboda, Němec, Dvorský, and Syllaba, along with Sis, were particularly outspoken. It was emphasized, especially by them, that certain matters should be kept secret from the Germans during times of potential political crises. They argued that considerations must be made whether, in their future work, they would have any secrets, secrets that Czech Freemasons might have to conceal, and whether there would be occasions when Czech Freemasons would prefer to hold discussions within their own completely confidential circle. They questioned whether such moments, when Czech Freemasons would consider it necessary to be alone among themselves, could also occur during normal times when hardly anything significant is happening.

Sis promptly responded to this question, saying, "There may come a time of defeat, temporary defeat, when Czech Freemasons will hide their secrets from the Germans, when maintaining secrecy will not only be a matter of Masonic honor but also a question of personal safety for Masonic lives... But then, it can only be a Czech closed Masonic circle, only Czech lodges that can pursue secret goals and employ secret means."

This clearly raised the question of effectively safeguarding the secrecy of Masonic activities.

The mention of "secret means" to be used for "secret goals" left no doubt in this regard. Nonetheless, the chairman of the newly established "Národ" Lodge believed that further clarification was necessary. During the inaugural ceremony held on November 6th, which took place for the first time according to Masonic ritual in the meeting room of the Národní klub, he stated in the lodge's programmatic speech:

"We hope that we will not have to resort to physical means to maintain our freedom. However, if it becomes necessary to use force to defend our national and human rights, in order to protect national freedom and independence, we must do so just as we did not hesitate to take any action in our struggles in the past, once a certain goal is proclaimed as the supreme law."

This statement made it clear that the lodge was prepared to take decisive action, if required, to defend national and human rights, even if it meant resorting to force. The mention of the "supreme law" indicated the lodge's commitment to a higher cause and the willingness to go to great lengths to protect it.

It was nothing more than a clear and unequivocal admission of potential terrorism! There was no need to refer to the statement made by representatives of French lodges at the meeting of Freemasons in 1861: "Charity is the consequence of our doctrines, but not the purpose of our gatherings."

Has there ever been a deeper acknowledgment of the principles of Scottish Freemasonry, which have entered history under the symbol of the dagger? Charity is a result of our teachings, but not the purpose of our meetings. Brother of the Rose Cross and Knight K and to the end.

The Grand Lodge in Rome deliberated in April 1919 on the request of the Prague Revolutionary Lodge "Národ" and granted its recognition to the newly established lodge. In the "Annuario 1918/19" yearbook, on page 39 under number 339, the entry "Praga-Nazionale 40" was published. This was the first public and official recognition of the "Národ" lodge within the framework of international Freemasonry.

On June 15, 1919, the Grand Lodge issued a charter to the Prague Lodge "Národ" and sent them the bylaws and master's degree certificates for its members. It was ordered that the "Národ" lodge be opened according to the ritual. The Freemasons in Prague then conducted new elections, where all officers and officials were confirmed in their positions, and they decided to open the lodge on October 28, 1919, the first anniversary of the revolution and the establishment of the Czechoslovak Republic. However, the inauguration had to be postponed to November 5, 1919, as the original date coincided with a national holiday that had its own extensive festive program, leaving little time for separate Masonic celebrations. One of the participants later wrote about his impressions of the final weeks of preparation, and we note them for their characteristic sentiment: "The brethren are obviously impatient. For us, the initiation of the rituals of the 'Národ' lodge is a historic moment. It seems to us as if we are introducing Czech Freemasonry into Bohemia with the task of safeguarding the nation's freedom and everything that the liberation revolution has achieved, and by concentrating men who were active in the underground revolution in Masonic lodges, continuing the work of October 28. But above all, we were all subjected to the mysterious enchantment of October's work, which Freemasonry had for us with its role in great moments of world history, in the grand development of England, in the French Revolution, in the struggles for freedom in all European states in the 19th century, especially the participation of Freemasonry in the fight against the Habsburgs."

During the preparations for the "Národ" Lodge, several high-degree Scottish Freemasons symbolically came to Prague to demonstrate the support of the entire Scottish Rite Freemasonry. They formed a lodge of the 4th degree (Perfect Master Lodge) according to the regulations. These individuals were primarily members of the French trade mission. The establishment of this lodge was approved by the "Supreme Council for Italy" and was announced in the official lodge yearbook for 1919/20. The newly established lodge had the task of assisting the Czech "Národ" Lodge in its initial work and preparing for the establishment of further high-degree lodges, which would be founded when Czech Freemasons reached higher degrees of the Scottish Rite. Its members had several consultations with the leaders of the "Národ" Lodge, which took place in the rooms of the Foreigners' Club on Slavonic Island. During these meetings, the Czechs also became acquainted with French Freemasons, including Freemason E. Basire, who held the 33rd degree of the Scottish Rite. Finally, the Yugoslav Grand Lodge also expressed great interest and actively supported the young Czech Freemasonry movement.

Meanwhile, a communication arrived from the "Supreme Council for Italy," in which Freemason František Sis was appointed as a member of the "Supreme Council of the Ancient and Accepted Scottish Rite" and elevated to the 33rd degree ("Sovereign Grand Inspector General") for his outstanding contributions to establishing Freemasonry in Bohemia. His 33rd-degree Masonic certificate was dated June 15, 1919. Additionally, Sis was appointed as the delegate of the "Supreme Council for Italy" in Prague. In a letter from the "Supreme Council," signed by the Grand Commander Raoul Palermi, Sis was hailed as the "founder of Scottish Rite Freemasonry in Czechoslovakia" and authorized to submit proposals to the "Supreme Council" regarding the premature conferral of higher degrees to Czech Freemasons for their exceptional merits. The following limits were set: 5 members for the 33rd degree (Grand Inspectors General), and 26 members for the 4th to 32nd degrees. Sis promptly submitted the relevant proposals, which Palermi approved in all cases in October.

In order to complete the ceremonial opening of the "Národ" Lodge, which was finally scheduled for November 5th, the Freemasons needed a suitable temple. They had their eye on several houses, particularly the house of the painter F. Engelmiiller. They considered it especially suitable for two reasons. Firstly, it was an old house and a former monastery, but its location on Hradčany (the Prague Castle) would symbolically underscore the Scottish Freemasonry's claim on Prague. Secondly, the Maffie had already considered this house during the World War; Dr. Rašín and sculptor Čapek had inspected it at that time, and the Maffie had intended to establish a secret printing press and weapons storage there.

Now, the Freemason Jindřich Čapek reopened negotiations with the aim of either acquiring the house through purchase or at least establishing a temple for the "Národ" Lodge there. However, the negotiations did not yield any results, so they had to agree to use the meeting room at the Národní klub (National Club) for the inaugural ceremony. The room was decorated and prepared according to the ritual regulations of the Freemasons. A foreign Freemason provided the master's gavel and a ceremonial carpet painted on paper, the master's chair and altar were set up, and the chairs were arranged in the proper order.

The ceremonial opening proceeded with a letter of greeting from the Supreme Council, which was represented by another special delegate alongside Sis. Following the introductory speeches by both delegates, the election resulted in the following appointments: Master of the Chair: Sis. His Deputy and First Overseer: Němec. Second Overseer: Dr. Syllaba. Speaker: Svoboda. Secretary: Dvorský. Tyler: Jindřich Čapek. Ritual Expert: V. Dyk. Ceremonial Officer: Ventura. House Administrator and Keeper of the Seal: Dr. Thon. Treasurer: Folprecht. Archivist: Borovička. Then a letter was read, confirming Sis's proposals for conferring higher ranks and degrees. The following appointments were included:

The 33rd Degree (Sovereign Grand Inspector General) was conferred upon:

1) Dr. Alois Rašín, one of the founders of the Maffie during the World War, who was sentenced to death for his activities

but received a pardon after a long period of imprisonment.
2) Dr. Přemysl Šámal, one of the leaders of the Maffie who played a significant role in underground intelligence operations during the World War, particularly in establishing connections with foreign entities.
3) Dr. Josef Scheiner, one of the leaders of the Maffie, who was imprisoned for years during the World War and maintained absolute silence.
4) Dr. Bohumil Němec, a university professor and one of the founding members of the resistance group "Národ" (Nation), as well as an advocate against activist policies during the World War.
5) Josef Svatopluk Machar, a poet and writer, who was a member of the Maffie and endured imprisonment during the World War while demonstrating "the spirit of Freemasonry throughout his literary work"...

The 32nd Degree (Prince of the Royal Secret) was conferred upon

Ing. Jan Dvořáček as a special trusted member of the Maffie.

Dr. L. Syllaba was promoted to the 31st Degree (Grand Inspector Inquisitor Commander).

Dr. Dvorský was promoted to the 30th Degree (Knight Kadosh).

> Dr. Svoboda,
> Jindřich Čapek,
> Viktor Dyk,
> J. Ventura,
> V. Pospíšil,
> Dr. Kapras,

and Dr. Thon were promoted to the 18th Degree (Knights of the Rose Cross).

Other members of the "Národ" lodge were promoted to the 9th Degree (Elect of the Nine) and the 4th Degree (Secret Master). With these mass appointments and promotions, the foundation was laid for Czech Freemasonry of the higher degrees. The aims of the high Scottish degrees, if revealed to those outside of Freemasonry, vary at each degree. For example, the 8th Degree emphasizes the brotherhood of humanity, the 9th Degree focuses on vengeance against ignorance, and the 26th Degree seeks the happiness of the soul, and so on. The Scottish Rite rituals incorporate certain props, such as the dagger as a symbol of vengeance against traitors, lamps to dispel darkness, and springs to quench thirst. The elevated Freemason is always asked the question of why they are "chosen," and the response is given that the death of Adonhiram must be avenged, and his murderers punished.

Rosicrucians and Knights Kadosh

A special ritual is prescribed for the initiation into the rank of Knights of the Rose Cross (18th Degree). The reception room is completely decorated in black, and the participating Freemasons sit in black attire with their gaze lowered in a sign of mourning. The candidate is welcomed with the words: "Brother, you come at a time that fills us with concerns and deep sorrow. The sanctuary of our tradition is in ruins; you see the wreckage that remains after this catastrophe. The temple of Freemasonry is shattered, the tools and pillars are broken. The star of truth has extinguished, the light of philosophy is obscured, and darkness and ignorance spread throughout the land.

The word has been lost (according to the "Great Constitutions" of the Ancient and Accepted Scottish Rite, revised by the Convent in Lausanne on September 22, 1875).

The acceptance concludes with the "discovery of the lost word." The introducing brother enters slowly with the candidate. The candidate's head is covered with a veil. They stand together in the west. All brothers rise, swords pointing towards the floor and with their right hand on their heart.

Question: Brothers, from where do you come?
Answer: Through the darkness of the grave, the valley of death, and the fire of suffering, we have sought the Word, and we believe that we have finally found it.

Question: Have you found the Word? How and in what manner?
Answer: It was when we were exhausted from the hopeless journey, unable to continue on the path, when our eyes no longer saw any destination on the path we had lost - when our ears heard no sound, and our lips could no longer utter a word - it was then that we, people in the desert of despair, fell without courage and utterly drained. Our state resembled a desperate mortal struggle, and before our eyes, the terrifying figure of death loomed. The hours passed slowly, silence surrounded us, night arrived with its horrors, and silently, eternity began to unfold its parchment, with not a single star visible in the dark sky...
As soon as we regained some strength, we recognized what awakened us; it was a voice emerging from the depths of our souls,

whispering words, while on the horizon, a barely visible glimmer appeared, like the first sign of dawn.

And thus spoke the voice: Arise, children of widows, and be strong! Though the light of the sun has extinguished and darkness reigns upon the earth and in your temples, humanity has dissolved in sorrow and tears. But this state has lasted for too long. Break the chains that still bind your spirits! (according to the "Grand Constitutions").

During the initiation into the degree of "Knight Kadosh" (30th degree), four rooms decorated in different colors are used.

The elevated one is informed that the legend of Adonhiram, which was told to him during his initiation as a master, is a mere fabrication. "During the initiation into this degree, it is still customary in many countries that the initiate, when reaching the highest rung of the ladder and hearing its explanation, suddenly falls to the ground, as the two side parts were drawn apart. It shows him how often a person, amidst the heights of glory, can fall to the ground. This sudden and unexpected fall is a symbol of the misfortune that can befall anyone, regardless of their knowledge or virtues." (according to the "Grand Constitutions").

The Czech Freemasons of the lodge "Národ" were exempted from all these ritual regulations when they were elevated to higher ranks during the inaugural meeting on November 5th, regarding the initiation of their own persons. The Supreme Council explicitly stated that it considers the activities of Maffie and the "Národ" group during the World War as a "proof," and therefore, the symbolic proof based on ritual order could be dispensed with. However, this did not mean that the Czech Freemasons of higher degrees renounced this ceremony for their future activities. On the contrary, it was expressly stipulated:

"It is understood that the lodge Národ regarded symbolic reception and the black chamber of contemplation as necessary rituals for initiation into the lodge because in ordinary times, in which we live, seekers are not fortunate enough to demonstrate Masonic qualities, moral courage, and character idealism through genuine deeds, and they can only submit to the symbolic proof of their Masonic character."

By a fundamental resolution of the lodge "Národ," it was stipulated that when new lodges of the Scottish Rite were established in Prague, only those individuals who had proven themselves in the activities of the Czech resistance during the World War should be initially admitted. The formation of additional lodges was planned to commence in the autumn of 1919. As a necessary preparation, the Freemasons of the lodge "Národ" compiled lists of individuals from all sectors of Czech public life, selecting those whom they deemed to be "ideological Freemasons" based on their spirit and character.

The selected individuals were then to be distributed among different lodges according to their professions and spiritual inclinations, creating harmonious centers in the lodges for various aspects of life, with the aim of representing the entirety of Czech life within the lodges. The existing lists of members from Maffie and the "Národ" group were supplemented with a multitude of personalities from the fields of science, cultural life, technology, and Czech national organizations.

After the ceremonial opening of the lodge "Národ," one of the main tasks was considered to lay the foundations for the establishment of additional lodges through weekly meetings, serving as "centers of Masonic work." Plans were discussed to establish lodges such as "Umění" (Art), which would bring together visual artists, musicians, and writers; "Dobrovský" for scientists; "Šafařík" to foster Czech and Slovak mutual understanding, also serving as a "Masonic gateway to Slovakia"; "Dílo" (Work) for engineers, industrialists, and practical professions; "28. října" as an association of Freemasons on a non-partisan basis; and "Purkyně" for the general goals of Freemasonry.

It was stipulated that the Freemasons of the lodge "Národ" in the rank of Master would be divided among the newly established lodges, and these lodges would be ritually constituted according to the existing statutes of the Grand Lodge of Rome.

Long discussions consistently emphasized the need for special caution in selecting individuals to engage with. In particular, Freemasons Syllaba, Emil Svoboda, and Dvorský persistently emphasized the stance of strict selection. According to known records, the following additional members were admitted to the lodge "Národ": Chotek, Krofta, Dr. Van'ček, Hanuš Jelínek, Dr. Vaněk, Dr. Politzer, Dr. Urban, Antal Stašek, Scheinpflug, Purkyně, Puc, Kotěra, Germář, Vlček, Švabinský, Štursa, Dr. Heller, Kadlec, Hora, Fr. Slavík, A. Procházka, Jakubec, Štěpánek, Rotnágl, Dr. Brabec.

According to the records, the Freemasons were to be divided into lodges as follows:

In the lodge "Dobrovský": Němec, Syllaba, Krofta, Hora, Kadlec, Kapras, Sommer, Jakubec, Petr, Slavík, Chotek, Babák. The following individuals were to be asked if they wanted to join the lodge: Drtina, Weigner, Pátá, Kümla, Hückl, Fridrich, Kaláb, Niederle, Domin.

In the lodge "Umění-Týn": Karel and Josef Čapek, Dyk, Svabinský, Štursa, Antal Stašek, Jelínek, Kotěra, Scheinpflug. The following individuals were to be asked: Dr. Guth, Křička, Španiel, Folkmann.

In the lodge "Dílo": Rašín, Pospíšil, Purkyně, Puc, Matys, Germář, Jareš, Schwarz, Pátá, Schustr, Klement.

In the lodge "28. října": Vaněk, Politzer, Nušl, Brabec, Rott, Bernášek, Gintl, Chytil. The following individuals were to be asked: Eršil, O. Černý, Sommer.

In the lodge "Šafařík": Vlček, Rotnágl, Hrušovský, Pilát, Stypa, Chotek, Folprecht, Bělohrádek, Táborský.

The proposal regarding the lodge "Purkyně" was to be postponed, while another lodge "Fügner" was to be established.

In its final session before Christmas 1919, the Supreme Council acknowledged the new lodges. The following individuals were elected as Masters of the Chair: Dr. Scheiner ("Fügner"), Dr. Syllaba ("Dobrovský"), Dr. Vlček ("Šafařík"), Kotěra ("Týn"), Puc ("Dílo"), and Ventura ("28. října").

At the same session, the Supreme Council approved the establishment of an independent Czechoslovak Grand Lodge. Its inaugural meeting was scheduled for December 29, 1919, and for this purpose, the assembly hall of the Národní klub on Celetná Street was ritually arranged again. There were two candidates for the office of Grand Master: the poet J. S.

Machar and the mayor of Sokol, Dr. Schemer. The Master of the "Národ" Lodge stated, "The goal was to build Freemasonry above political parties, and in that regard, my own person would be too politically exposed. On the other hand, both Machar and Schemer were highly suitable candidates. Machar, one of President T. G. Masaryk's closest friends, was particularly praised for his writings, some of which had explicit Masonic elements."

On December 29, 1919, at 10 o'clock in the morning, the inaugural meeting of the Czechoslovak Grand Lodge took place. The election had the following outcome:

> Grand Master: Machar
> Deputy Grand Master: Scheiner
> First Grand Overseer: Sis
> Second Grand Overseer: Kotěra
> Grand Speaker: Svoboda
> Grand Secretary: Dvorský
> Deputy Grand Secretary: Germář
> Grand Treasurer: Kapras
> Grand Controller: Ventura
> First Grand Expert: Němec
> Deputies: Vlček, Syllaba
> Grand Ceremonialist: Dyk
> Grand Almoner: Pís

Grand Tyler: Urban

The financial commission consisted of Kapras, Puc, and Ventura.

With this, the circles of the Maffie and the "Národ" group joined high-degree Freemasonry, not only personally but also found an organizational basis in the "Czechoslovak National Grand Lodge." By electing Machar as the Grand Master, the leadership of the new Grand Lodge was entrusted to a successful member of the Maffie during the time of World War I. At the beginning of the war, Machar lived in Vienna as a poet and managed to obtain important documents through Čech Kovanda, a servant of Austrian Interior Minister Heinold. Heinold would bring home files for daily reading, which his servant regularly copied for Machar.

Beneš evaluates this performance with the following words: "It gave us a unique insight into the plans, goals, views, and political methods of the Vienna government, Stürgkh, Heinold, Thun, and it rendered us great service in our work" (Beneš, p. 21). Machar also managed to obtain information about the political intentions of the state leadership, such as the military command's actions against the South Slavic Sokol, enabling him to timely warn those affected. He even obtained a copy of a telegram from the Austro-Hungarian embassy in Rome regarding Masaryk's planned return from Italy to Bohemia, which alerted him to the danger and allowed him to warn Masaryk in time, resulting in his staying abroad. Machar's activities led to "matters discussed in the Viennese Council of Ministers appearing in the allied press four or five days later. On other occasions, details from Vienna appeared in the 'Nation Tchèque,' which revealed, as we discovered, the agitation in Viennese government and police circles, as it indicated the good organization and swift communication between Vienna, Prague, and Switzerland, thus demonstrating the strength of Czech conspiratorial action" (Beneš, pp. 48/49). Only after several years did the state authorities take notice of Machar and take action.

Once the Grand Lodge was established, there was only one task remaining, which was to secure international recognition for it on all sides to ensure its undeniable sovereignty. On January 12, 1920, the Czech high-degree Freemasons, if they were in the 33rd degree, formed a provisional "Supreme Council" in

Prague, with Sis appointed as its chairman. His task was to negotiate the recognition of Czechoslovak Freemasonry by all the organizations of Freemasons represented in the world association of the Ancient and Accepted Scottish Rite (known as the "Lausanne Confederation").

The "Supreme Council for Italy" in Rome was willing to provide a guarantee and recommendation as the founder on behalf of the Czech Grand Lodge before the global Freemasonry community. When Sis went to Paris for an extended stay in 1920, he immediately visited the General Secretariat of the "Supreme Council for France" at 8 rue Puteaux and requested that the "Supreme Council for France" recognize the "Supreme Council" in Prague. Through Chaillé, Sis came into contact with the Sovereign Grand Commander of the "Supreme Council for France," R. Raymond, who also served as the Grand Secretary.

Sis joyfully reported on this: "Upon my arrival, the Grand Chancellor delivered a brief speech about Czechoslovakia, expressing his pleasure that Scottish Rite Freemasonry, according to its international mission, is expanding its influence in Czechoslovakia. In response, I mentioned the emergence of Czech Freemasonry in the struggle for national freedom and the efforts to build an independent state. My speech was warmly received by the members of the Supreme Council, and both the Sovereign Commander and the Grand Chancellor responded to it with heartfelt enthusiasm."

However, shortly after, Sis had to write to Dr. Dvorský, the Grand Secretary in Prague, stating that first, evidence must be provided that Czech Freemasonry was founded properly and precisely according to the statutes. The Supreme Council for Italy, despite pleas from Prague, did not lift a finger to issue a more detailed guarantee statement. At that time, there was some loosening of Masonic relations between Prague and Rome due to various personal conflicts.

It was only after two years that the Supreme Council for Italy began to act in the desired sense. This occurred during the third International Conference of Delegates of the Supreme Councils of the Ancient and Accepted Scottish Rite, held in Lausanne from March 29 to June 3, 1922. After the Czech Freemasonry gained general recognition and its delegate gave a speech, the Grand Chancellor R. Chaillé embraced the Czech speaker on behalf of French Freemasonry and congratulated him, according to preserved reports, on the fact that "Czech national Freemasonry had the most Masonic origins in the light of freedom and equality, surrounded by a history imbued with Masonic ideals, and wished Czech Freemasonry a future befitting its past."

Shortly thereafter, fascist Italy dealt a devastating blow to the seeds of Freemasonry within its borders, leading to the end of the Supreme Council for Italy and the Italian National Grand Lodge. In their place, the Yugoslav Grand Orient had to bring light to the Czechoslovak National Grand Lodge. Thus, recognition of Czech Freemasonry was achieved at the last possible moment. If the guarantors in Rome had fallen earlier, it might not have been possible a year later. Despite all the friction of recent years, Prague let out a loud lament and declared itself forever grateful and loyal to the founders. It was during that time that the funeral song of Czech Freemasons said, among other things:

"Therefore, in the book of the history of Czechoslovak Freemasonry of the Ancient and

Accepted Scottish Rite, the deeds of Italian Freemasonry will be recorded in this light. The great power of Italian Freemasonry has come to an end. In the new order introduced by fascism in all aspects of Italian life, there was no place for the association of free men whose free spirit is governed by nothing other than internal conviction and personal will. What remains of Italian Freemasonry are only scattered remnants of a great spiritual domain, but there is also hope and faith that freedom and the power of the spirit can only be enslaved temporarily."

Now, Paris took the place of Rome for Czech lodges. However, the relations between Czech and French Freemasonry had been very limited so far, mainly confined to Sis's personal contacts in Paris. In 1922, a certain unification of the two Czech Freemasonry groups, the "Czechoslovak National Grand Lodge," the "Supreme Council for Czechoslovakia," and the "J. A. Comenius" Lodge, which had previously existed in isolation and worked under the "French Grand Orient," was achieved. On May 16, 1922, the lodge decided to join the "Czechoslovak National Grand Lodge." Through a letter from the "French Grand Orient" dated June 14, 1922, it was released from French obedience. By joining the Czech Freemasonry organizations, they received the closest connection to France as a dowry. For none other than President Besnard of the "French Grand Orient" had brought light to the "J. A. Comenius" Lodge in 1919.

Thus, the doors were opened to the French model of "socialist spirit" with all its political consequences of clear left-wing orientation. The connection between Freemasons in Prague and Paris grew closer and manifested through mutual visits and honors. For instance, the later Czech Grand Secretary Leo Schwarz and Freemason R. J. Vonka were awarded the title of Honorary Masters of the "Shakespeare" Lodge in Paris during their visit. Similarly, both former presidents of the "French Grand Orient," Freemason senators Brenier and Wellhof, visited Prague and were honored with honorary membership. The French-Freemasonic thinking, complemented by the Brother of the 33rd Degree Edvard Beneš, contributed to the ultimate fusion, which then, albeit not always visibly, exerted unfavorable influences on Czech politics for the next two decades.

Thus, the Czech Maffia, a group of Czech resistance fighters during World War I, finally ended up in the clutches of the "French Grand Orient." However, the bearers of the Czech resistance during World War I had definitively sold themselves to the Western world, which was equally ideologically and morally decadent. This Western world then made Czech nationalism of the following decades immune to any healthy movement and burdened it with that mortgage, the political interests of which became clearly evident in the recent past.

"The Light" in the Sokol Organization

Those who had the opportunity in recent years to visit the Sokol museum in Prague's Tyrš House, far from the spirit of mass visits and stereotypical speeches of guides, and engage in contemplation, were probably disappointed by the musty atmosphere that pervades it. Even there, where the latest events in the history of Sokol are depicted in paintings and documents, such as the last gathering in 1938, one would look astonished at those busts and exhibition pieces, which are arranged in a peculiar manner on the walls and door frames even in this most recent exhibition space. A keen observer involuntarily senses a suffocating, almost Masonic impression. And with attentive observation, it is soon possible to identify objects and symbols that confirm this overall impression, even in detail, such as the Masonic hammer used by Czech Sokols in America. These associations raise questions for a more in-depth examination when assessing this historically significant and largest Czech physical education association.

An incorrect interpretation may lead some German observers to such conjectures based on minor details and generalize them. However, there is a whole range of relevant Czech and foreign voices that, mostly in recent years, arrive at similar conclusions. A reliable witness, even for Czechs, can be Wacìaw Filochowski, a Pole who wrote a book titled "Cierpkie Pobratymstvo" (Bitter Brotherhood) after a longer stay in Czechoslovakia in 1935 and 1937, which was published in Warsaw during the Czechoslovak crisis of 1938. The author sees the former republic, with its fragility as an unnatural nation-state, and writes what could be seen as a political obituary. The book is written in a journalistic and admonishing style, but its political subjectivity is purely Polish. The author looks through "polluted Polish windows" towards Teschen and particularly longingly writes about Slovakia.

However, in one aspect, Filochowski's perspective must be acknowledged as clear and unbiased, particularly in his uncompromising discussion of the Jewish question and Freemasons in his concluding chapter. Furthermore, the Polish author does not forget to mention events and observations he made during his stay in Prague with the Sokol organization. In several parts of his book, he points out how senior Sokol officials are involved in Freemasonry, and on one of the last pages of his work, he reaches the overall conclusion that "an organization of such great significance as Sokol is controlled by Freemasonry." Freemasons from the leadership of this organization worked to admit Jews as members but excluded uncompromisingly nationalistic individuals. Freemasons now regard members of the Czechoslovak Sokol as "Freemasons in profane life" (Filochowski, p. 301).

It would be incorrect to interpret this judgment as an eruption of Masonic obligations of individual members of the Sokol leadership or as a grotesque phenomenon of recent years. As early as 1911, the Petrograd magazine "Kolokol" paid attention to the Czech Sokol community and its leading representatives. It concluded that "the majority of intelligent Sokol members were radicalized by the influence of Jewish-Freemasonic press and pseudo-professors à la Masaryk." The magazine explicitly mentioned the name of the then Sokol chairman, Dr. Scheiner, without specifying whether it personally regarded him as a Freemason.

The question of whether Scheiner was already a Czech Freemason, like many others, before World War I and the establishment of independent Czech lodges, is still controversial. If that were the case, then Scheiner's secret trip to the Russian Foreign Minister Sazonov (who was a member of an English lodge) shortly before the outbreak of the war in February 1914, known today, would take on a completely new light. In the "International Lexicon of Freemasons" by Lennhof-Posener, the long-serving chairman of Sokol, Dr. Scheiner, has been praised so far only as a post-war Freemason. However, the evidence used for this work at least suggests that Scheiner was already involved in the conspiracy during World War I in František Sís's plans to establish a lodge and that he was admitted to Scottish Freemasonry immediately after the war through the Italian Grand Lodge, via the National Grand Lodge.

Although it is still not entirely clear how the relationship between Sokol and Freemasonry began, it can be reasonably judged that members of the lodges began to penetrate Sokol on a large scale only when the Czechs had their own state. This development is closely associated with the names of those two Sokol personalities who decisively shaped its destiny for almost five decades: the aforementioned Dr. Josef Scheiner and Dr. Jindřich Vaníček. Both were disciples and young collaborators of Sokol's founder, Miroslav Tyrš. They wrote important articles for the Czech gymnastics movement and served as chairman and deputy chairman of the Sokol Association, founded in 1889, until World War I. Scheiner was a prominent Sokol politician both at home and abroad, recognized as the father of the Pan-Slavic Sokol Union and one of the main Czech speakers at Slavic congresses before the war. Vaníček excelled as a gymnastics tactician and organizer of Sokol gatherings and gymnastics trips abroad.

Scheiner participated in the memorable Maffie meeting during Christmas 1914, where the establishment of a Czech lodge was discussed. Immediately after the war, he attained the rank of Master in the "Loggia Nazionale Oriente di Roma," and in 1919, he was awarded the 33rd degree of the Ancient and Accepted Scottish Rite Freemasonry. Vaníček was admitted to the lodge "Národ" (Nation) as a seeking brother in the first group. With Scheiner and Vaníček, who assumed the positions of chairman, deputy chairman, and chief after the revolution in the restored Sokol Association, the groundwork was laid for the penetration of Freemasons into the entire gymnastics movement. During the ceremonial opening of the lodge "Národ" on November 5, 1919, it was resolved that the first lists of lodge candidates should primarily include personalities from Sokol. Simultaneously, a plan was established to establish the "Fügner" lodge as the "Sokol lodge." The founding members designated were Sokol members and Freemasons Scheiner, Vaníček, Urban, Šámal, Dvorský, Jindřich Čapek, Heller, Jeřábek, Štěpánek, and Obešlo.

After the establishment of the Sokol lodge, the Supreme Council for Italy in Rome quickly approved its creation. Scheiner was elected as the Master of the Chair, thus simultaneously leading both the Sokol Association and the Sokol Lodge. This is a telling sign of the beginning of intertwining. When the National Grand Lodge of Czechoslovakia was founded on December 29, 1919, Scheiner was elected as its Deputy Grand Master, narrowly missing out on becoming the Grand Master instead of poet Machar. Scheiner was also a member of the Supreme Council of the Ancient and Accepted Scottish Rite Freemasonry in Prague, which was formed on January 12, 1920. It was probably at his instigation that the Czech lodges at that time decided to embark on a stronger recruitment campaign within the Sokol circles.

Scheiner soon found enthusiastic fighters for the fusion of Sokol with Freemasonry. One of the most zealous was the then associate professor and later professor at the Czech Technical University in Prague, Dr. Ing. Rudolf Bárta. This Sokol-Freemason presented a plan in the lodge "Dílo" (Work) that was of great significance for the expansion of Freemasonry and was also published in the Journal of Czech Freemasonry (Year X, Issue 2). It culminated in the following five theses:

1) Sokol provides a suitable ground for the development of Masonic activities because they share common ideological content.
2) Both associations differ only in their establishment, working methods, and, of course, the idea itself. Sokol is democratic, while Freemasonry is already aristocratic in its establishment, and the principle of authority is applied at every step, with rights granted gradually through "salary increases." Sokol is built from the bottom up, while

Freemasonry is structured from the top down. Sokol is a public organization, while Freemasonry is an exclusive order. Sokol's work is based on voluntary discipline, with transparent, purposeful, and systematic working methods, aiming for the shortest path to the goal. Sokol is a living association that complements and adapts its working methods according to the demands of the times. Freemasonry, on the other hand, envelops its work in rituals accompanied by symbols that speak to our soul, not through modern practical means but in a manner similar to religion. Masonic working methods have remained essentially unchanged for over a century. There is also a significant difference in the idea itself between Sokol and Freemasonry.

3) Both associations are progressive. In both, there are people of goodwill who seek to influence their surroundings and engage in education. While we don't know for certain whether Tyrš was a Freemason behind the scenes, it is likely that he was, albeit without a visible presence.

4) The significance of cooperation: Freemasonry, which is and will remain a narrower closed circle, would gain the opportunity for broader influence through closer cooperation with Sokol. It is a matter of implementing the principles that Freemasonry has in mind more assertively, which are already a part of Sokol's activities. Freemasonry would thus establish greater contact with broader sections of the nation at a time when Freemasons elsewhere are losing ground. This would benefit our consolidation.
5) Cooperation could be organized in each "Orient," primarily in Prague. An inter-lodge group would be formed consisting of Freemasons who are also Sokol members. Their task would be to bring as many capable Sokol officials into the lodges as possible and vice versa, to promote Masonic principles within Sokol and vice versa. We already have several brothers who are both dedicated Sokol members and officials. If their ranks were to increase, it would already have positive results and effects.

The Freemasonic decay continued systematically. Within the Sokol organization, a deliberate personal policy was initiated by Freemasons. Freemasons were appointed to important positions, both in the central leadership and in leading positions within Sokol's regional divisions. The Sokol ideology was appropriated by Freemasonry and interpreted or distorted to align with their own agenda. Let us take a look at the promotional publication "Československý Sokol" (Czechoslovak Sokol), published in 1932 by the Sokol Publishing House in Prague. In the chapter on the Sokol ideology, we find these significant words:

"The motto of the French Revolution, 'Equality, Liberty, Fraternity,' discipline, and morality form the first main component of the Sokol ideology. The second constant element is the constant endeavor of Sokol to embrace the entire Czechoslovak nation within the conscious, progressive spirit of the Sokol ideology. In this sense, Sokol understands its obligations towards the nation and fulfills them. It simultaneously works towards achieving the humanitarian ideal, which aims to establish a fraternal relationship among all cultural nations, working together as equal partners to realize the universal ideals of humanity."

In the publication "Výhled" (Outlook) on page 47, we find the same ideas when Sokol declares its commitment to "a culture that is guided by the principles of brotherhood, humanity, justice, and truth." These beautifully sounding words were nothing other than the ideology of French Freemasons. There was no longer any need to conceal the true goals and hidden motives. By 1932, Freemasons had firmly established themselves in all positions within the Sokol organization, making it unlikely for any significant resistance to arise from the membership.

Let us briefly consider the outcomes of Freemasonic personal politics within Sokol over the past two decades! After the World War, Dr. Josef Scheiner, a Freemason of the 33rd degree, became the President of the Sokol organization. Freemasonry did not grant this highest position of presidency for twenty years until Sokol's activities were halted.

After Scheiner's death in 1932, Dr. Stanislav Bukovský, a respected Freemason and a physician in Prague, took his place as the

Deputy President of Sokol. Since 1921, Bukovský had been a member of the male leadership, overseeing the Sokol department for medical examinations, which he had initiated. In 1925, he was appointed to the presidency of the Sokol organization, and in 1931, he became Scheiner's deputy. However, when Bukovský became intolerable due to his explicitly left-wing orientation after the Czech collapse, his position was taken by Josef Truhlář from Poděbrady, a retired senior official, as the Sokol president after the establishment of the Protectorate.

It was widely claimed in the Czech press in recent years that Truhlář was also a Freemason. Since 1930, he had been the leader of the Tyrš Sokol District in Kolín, and he became a member of the Sokol Education Department, which was a special area of influence for Freemasons. He then served as the third deputy under Scheiner and the first under Bukovský until succeeding Bukovský in the leadership of the Czech Sokol organization.

Next to the president of the Sokol organization, the position of the leader (náčelník) was considered the most important because the leader oversaw the training of active members.

It is understandable that this significant position within Sokol was also dominated by Freemasons and remained so from the World War until the present day. For approximately 15 years until 1933, Dr. Jindřich Vaniček, already known as a Freemason, held this role. When Sokol experienced a significant increase in membership after the war and the duties of the leader grew, a special training council was formed. Freemasons Dr. Klinger, Dr. Bukovský, and Dr. Pechlát were immediately appointed to this council.

Later, Dr. Miroslav Klinger, a senior official in the Ministry of Health, replaced the aging leader Dr. Vaniček as the deputy. When Vaniček retired due to old age two years later, Klinger took his place, which satisfied the Freemason reservation. However, Klinger's five-year tenure as leader and his prominent affiliation with Beneš's political party burdened him so heavily that, after the establishment of the Protectorate, he had to yield to public pressure, just like Sokol President Bukovský. His position was then succeeded by his former deputy, Dr. Augustin Pechlát, who was also an active member of the Prague lodge and a speaker for the "League for Human Rights."

Pechlát served as the leader of Sokol until its dissolution, and on September 30, 1941, he was sentenced to death by a military court in Prague for his involvement with a resistance group.

After the president and leader, the third most important function within Sokol was held by the educators and trainers, who had their leadership and management within the educational department of the Czech Sokol organization. The role of a Sokol educator was similar to that of a "Dietwart" in the Sudeten German Turner tradition, focusing on intellectual development and spiritual orientation towards the idea of national physical education. To a critical observer, it becomes evident how deeply Freemasons penetrated the field of education within Sokol.

Although detailed personal compositions of the educational department's presidency are recorded in Sokol's yearbooks only from 1928/29 onwards, a clear picture emerges within those twelve years. Freemasons such as Ladislav Jandásek, a regional school inspector in Brno, Dr. J. Kozák and Dr. Emil Svoboda, both university professors in Prague, as well as Truhlář, were already shaking hands as members of the presidency in 1928. In 1937, they were joined by Freemason Dr. Augustin Pechlát, who would later become the leader of the Czech Sokol organization. In 1928, Freemason Dr. Karel Sirotek, a specialized teacher in Brno, was also part of the educational department. Finally, in 1940, Professor Dr. Ing. Bárta from the "Dílo" lodge, the originator of the plan for Freemasons to infiltrate Sokol, appeared as an educational official in the presidency. Similar patterns were observed in the educational departments of regional branches, where Freemasons had also taken control of the educational activities.

It is no wonder, then, that even the representative and advisory board of the Sokol leadership, the Central Presidency of the Czech Sokol Organization (ČOS), was increasingly dominated by Freemasonry in the last twenty years. In addition to the previously mentioned leading officials, men such as the later Grand Master of the National Grand Lodge of Czechoslovakia, University Professor Dr. Karel Weigner from Prague, Professor Dr. Bohumil Kladivo from Brno University of Technology, National Democrat Dr. Jan Stolz, Brno University Professor Dr. Vladimír Groh, and others appeared within this body.

The Presidential Council, the highest advisory body of the Sokol leadership, was even more under the control of Freemasonry. Freemasons also held sway in the financial department, the economic department, and, above all, in crucial positions concerning personal politics within the Honorary Council and the Educational Department, the Exclusion Committee, and the Personal Department of the ČOS. It goes without saying that the leadership of the American branch of Sokol was shaped exclusively by Freemasons.

It would be tiresome for readers if we were to list similar examples of this decay within the leadership of regional branches of Sokol, which were no different from the central leadership. Let's take, for example, a single rural Sokol region, the "Jana Máchala" region in Brno, as a representative example of the state of affairs in rural areas. Freemasons held the following positions there:

- President: Dr. Bohumil Kladivo, Professor at the Technical University in Brno.
- First Deputy: Čeněk Krejčí, specialized teacher in Hrušovan near Brno.

- Educator: Karel Sirotek, specialized teacher in Brno (also a member of the "Most" lodge in Brno).
- Deputy Educator: Ladislav Jandásek, regional school inspector in Brno (also an official of the "Cestou světla" lodge).
- Treasurer: Dr. Zdeněk Krejčí, bank officer in Brno.
- Social Affairs Officer: Dr. Josef Kudela, gymnasium director in Brno.

This pattern, with Freemasons controlling key positions, was largely repeated in other Sokol regions as well.

The control of the Sokol organization by Freemasons extended beyond the Czech Sokol organization and reached into the pan-Slavic Sokol movement. The highest international organization of all national Sokol organizations was the Slavic Sokol Federation, which had a pre-war tradition. Even here, the influence of Czech Freemasons was decisive. The leader of the Sokol organization and a Freemason, Dr. Miroslav Klinger, served as the chief, and the treasurer was Josef Truhlář. The president of the Slavic Sokol Federation was Czech Freemason Dr. Stanislav Bukovský, represented by Yugoslav Freemason Engelbert

Gangi from Ljubljana. At the international Sokol gathering, Sokol members who were Freemasons from various countries met in Sokol uniforms to declare their commitment to the "universal goals of Sokol."

In the autumn of 1938, when attempts were made to reorganize Czech national life after the collapse of Czechoslovak foreign policy in the Second Republic, which soon proved to be only political gestures due to the presence of many Freemasons in the government, it seemed that a discussion about Freemasonry and Sokol would also arise. Therefore, the leadership of the Czech Sokol organization felt compelled to take a stance on Freemasonry. According to an old tactic, the issue was handled in a very vague manner. The article published in the Sokol's press organ "Věstník" on October 23, 1938, attempted to create an innocent impression by linking Freemasonry to medieval guilds and by diverting attention from the dangerous problem through incorrect numbers. Under public pressure, Sokol members with Masonic connections, Bukovský and Klinger, had to step down, but they were immediately replaced by brethren from the lodges, Schauer and Pechlát. Sokol took a decisive stance in support of

Freemasonry when it clearly stated in the mentioned issue of "Věstník" (Bulletin):

"Domestic malicious propaganda and external incitement, influenced by it, suddenly found culprits for the disintegration of the state: Freemasons and Rotarians, among whom Sokol was also implicated in Slovakia and partly in the historical lands. Sokol is allegedly led by Freemasons, and because Freemasons are deemed guilty, Sokol is also accused of the unfortunate policies that led to the downfall.

Even though it is nothing but malice and an incurable ignorance (despite having enough books on Freemasonry in Czech), that have concocted this trio and accused them of various intrigues, it should be noted that for Sokol, it is by no means a bad company. After all, members of Freemasonry and Rotary have been working diligently, selflessly, and tirelessly for the same goals for twenty years in their global associations, just as Sokol fulfills them at home and abroad: to elevate their nation through wisdom, strength, and beauty."

Looking back at the twenty years of post-war history of the Czech Sokol, it entered a new period as a well-organized and tested Czech physical education union, whose growth, development, and impact in the decades before the world war could not escape political attention. Illuminated by the glory of Czech resistance during the world war, it had favorable conditions in the young Czech state and all the possibilities for growth to become a great educational organization for Czech physical education in a national spirit.

The outward signs of growth seemed to indicate that these opportunities for success would be utilized. By 1920, the Czech Sokol boasted approximately 325,000 members. Its peak came in 1937, when it had well over 700,000 members, including youth and children. This meant that on average, every ninth Czech was a member of the large Sokol family.

However, this unprecedented growth in numbers did not correspond to its internal development. Sokol was losing its persuasive power from its earlier years and succumbing to influences that increasingly alienated it from its national responsibilities. Just as the growing influence of Judaism threatened to disintegrate the Czech Sokol from within, the infiltration of freemasonry alienated its leading positions in the central administration and lower organizational units, leading to an intellectual decay from the top. Even Sokol, as a significant Czech educational union, was immunized against national impulses for a true national revival, which was necessary in the coming years but did not occur at that critical moment in Czech history.

It is more than characteristic that on May 30, 1932, the Grand Secretary of the National Grand Lodge of Czechoslovakia sent an official letter to the Grand Secretary of the National Grand Lodge of France, inviting Freemasons to participate in the Sokol rally and to circulate it among them. The connection between Sokol, Freemasonry, and Judaism became even clearer a few years later when, in September 1935, just two letters from the Freemason-infected "Jack London Club" in Prague and the "Jewish Party in Czechoslovakia" sparked passionate discussions within Czech Sokol about the possible boycott of the Berlin Olympics. The Sokol leadership quickly requested the opinion of Brother Ladislav Jandásek, a member of both Sokol and the lodge in Brno, on this matter on September 30, 1935.

A few months later, a new wave of protest letters from the "Association of Jewish Workers in Prague," the "Syndicate of Working Women's Intelligence in Czechoslovakia," the Czech "League Against Anti-Semitism," and others provoked the Central Leadership of the Czechoslovak Sokol to initiate, on March 25, 1936, a "very cautious" discussion among gymnastics organizations in other countries about participating in the Olympics or renouncing it. Subsequently, relevant inquiries were immediately sent out "in light of changing political circumstances and with regard to the possibility of further complications in international relations" to Paris, Brussels, Aarau in Switzerland, Warsaw, Sofia, and Belgrade. The response from the Bulgarian Junaks contained a clear and unequivocal declaration that they would go to Berlin. Others also stated that they would participate through the Olympic committees of their respective countries. In a delusional belief that they were acting as vanguard fighters against anti-Semitism, Czech Sokol completely isolated itself and was abandoned even by its former friends.

According to the observer's perspective, the collapse of Czech Sokol can be considered a national tragedy or a deserved fate. However, one thing is certain: the end was inevitable.

While on the other side of the Czech national borders, a process of purification and national awakening in physical education was taking place, Czech Sokol was deteriorating internally and intellectually despite its external splendor. Eventually, it met its inescapable fate when it became entangled in the service of world Jewry and Freemasonry.

The Temple of Youth[ii]

The infiltration of foreign Masonic elements into Sokol and their influence on the Sokol movement, when viewed from a Czech nationalist perspective, meant that the most promising groups of Czech men, raised in the spirit of service and selflessness, were ideologically subdued. However, from a historical perspective, it is even more significant that through this process of intellectual decay, Sokol ceased to be a true agent of national youth education. Considering the development of the past twenty years, which Czech organization had better opportunities to intellectually guide Czech youth than Sokol? With its own youth groups and by extensively involving Czech schoolchildren of both sexes in its gymnastic halls, Sokol had unique foundations to exert a decisive educational influence on the growing Czech youth beyond school and family.

There is no recent data on the exact number of young people Sokol has attracted. Nevertheless, official reports for the year 1937 provide an approximate picture of Sokol's educational influence on Czech youth. On Defense Day, April 18, 1937, which was held in 3,000 locations throughout the republic, 42,785 male youth (93.43%), 44,597 female youth (98.38%), 94,362 male students (73.72%), and 101,036 female students (65.79%) participated actively. These figures primarily indicate that over 280,000 young Sokol members actively participated in the Defense Day event in 1937. Furthermore, considering the total number of Sokol youth at that time, they clearly demonstrate the dedication of Sokol's youth, which was evidenced by the relatively high attendance at this special general rehearsal. What prospects for educational success could a morally upright association have alongside its own physical education work and educational activities?

The disappointment of Sokol, in terms of truly national youth education transcending chauvinistic tendencies, was even more significant because the Czechs did not have other similarly comprehensive organizations during their own statehood. Other physical education organizations, such as "Orel," were defined by their religious affiliation, while the DTJ (German Workers' Gymnastics and Sports Union) had a class-based orientation. Additionally, almost all Czech political parties created their own youth organizations, which, for the most part, did not contribute significantly to youth education, except for some exceptional cases among young agrarianists. Their members were not much more than readers of politically controlled youth magazines and political agitators in secondary schools, whose focus revolved around the question of securing positions through party politics after finishing school, without much competition. In their national saturation, resembling the German situation before World War I, and the national void following the revolution, the Czech nation lost many active young people to communism.

The souls and consciences of nations react differently to the same political realities, and the post-war Czech youth responded differently from the pre-war German youth. There was no spiritual upliftment in the sense of the German youth movement during the Wilhelmian era among the Czechs. After World War I, a movement emerged among Czech youth that advocated the recreational use of free time in nature, which peaked in the early decades of the previous century and continues to this day: tramping. However, tramping did not develop into anything more significant than a "popular sport, such as stamp collecting and weekend trips." A keen observer of Czech youth described it years ago as follows:

"The majority of average Czech youth escapes from harsh reality into romantic daydreaming of tramping. These uninvolved young people flock in groups to the forests on weekends, paddle canoes to their cabins, sit by small campfires at night with their Indian-style dressed girls, and sing songs about wandering life and sentimental romanticism accompanied by balalaikas and Hawaiian guitars" (Fritz Leif in the magazine "Völkische Stimmen," Prague, February 1937).

Indeed, it was not a political uplift or a revolutionary beginning of youth yearning for revival, but rather a call from the youth, tired of the big city, for the American cowboys, a modern outlet for world-weariness and the mal du siècle.[iii]

What else remained? There was still a numerically significant, organized Czech youth who, far from party disputes, without confessional or class prejudices, sought to create new forms for the acquisition and education of Czech youth. These Junaks (Scouts) were fond of uniforms, external discipline, selection, and, disregarding the short period of romantically diluted wandering and camp life at the end of the first decade after World War I, militarily disciplined behavior. They emerged from the first scout groups before the war and established themselves as the organization "Junak - Czech Scout" in 1914.

As we saw during the depiction of the domestic resistance during World War I, they became young leaders of Czech illegality. In the last ten years, the Junak movement became increasingly politicized and integrated into pre-military education. At the same time, it succeeded in becoming the largest central movement of Czech youth, especially those interested in sports and physical education, and other Czech scout groups of Catholic, agrarian, and liberal orientation continued to join them. There was no doubt that the most active part of Czech youth was gathering here.

Thus, Junak became the leading educational association for the emerging Czech generation.

In recent years, "Junák" has brought together around 40,000 boys and girls. They were divided by gender and organized into three age groups. Children aged eight to twelve were called "vlčata" (wolf cubs), those aged twelve to eighteen were called "junáci" (scouts), and those over eighteen were called Old Skauti (senior scouts). Within these age groups, the smallest unit was a družina (pack) consisting of 6 to 12 members, followed by an oddíl (troop), sbor (group), okres (district), and župa (region). At the head of the entire organization was the so-called hlavní stan (headquarters) located in Prague.

The hlavní stan consisted of the central council responsible for administrative matters and the náčelnictvo (leadership) entrusted with the educational leadership of Junák. Like any nationally organized youth movement, the Czech Junák scouts were ideologically rooted in the traditional dependence on the Anglo-Saxon world. The founder and long-time leader of Junák, Professor A. B. Svojsík, studied Baden-Powell's "Boy Scout" movement during an extended stay in England.

Upon his return in 1912, he established the first Czech scout družina in Prague based on its model. In his speech on November 28, 1918, Svojsík mentioned how the organization came into existence. When he presented the principles of the English "scouting for boys" to

Professor Masaryk, who was already ideologically oriented towards the West, Masaryk greeted Svojsík with the words, "Ah, I already know this work, and I will support scouting in our country wherever I have influence."

After the World War, Masaryk, as the President of the Republic, also became the protector and a special supporter of the Junák Association. In the practical activities of the following two decades and during the World Jamborees (large youth camps), the Czech scouts always strived to maintain close connections with English and American scouts and eventually earned recognition to host the World Jamboree in Prague in 1941, which did not take place due to the events unfolding at the time.

Even though the general sentiment among the Czechs turned against England and France after the collapse in the autumn of 1938, Junák, as the first Czech organization, expressed its commitment to maintaining the old friendship openly in its magazine "Činovník" (Volume XIX, Issue 2). It was not just an external organizational connection between the Czech Junák and the Anglo-Saxon scout movement that determined their dependence on England, despite disappointments in foreign politics caused by Western powers. There was a deeper bond that existed unnoticed by outsiders and individual scouts within the Czech scout leadership.

Svojsík, in his post-revolution speech, spoke of the "powerful enchantment" inherent in scout education and praised the pre-war work of Czech scouts in promoting the idea of "peace scouts, from whom the rebirth of humanity is expected" and that it is "an idea that already subtly moved beneath the surface of general thought and is the guiding star of the future organization of the world." These statements by Svojsík about the renewal of humanity and the "universal" world view already hinted at the true background and its reasons. However, it was Dr. Zdenko Štekl, the regional leader of Junák in Moravia, who explicitly stated what was only implied in the words of the founder and long-time leader of the Czech scout movement. In March 1938, Štekl privately published a work titled "Skaut - Malý zednář" ("Scout - A Little Freemason") in 300 copies, which was circulated confidentially among a select group of Junák leaders. This richly illustrated booklet, adorned with various symbols on the margins, had a meaningful title that indicated the intentions of its author and its intended audience. The fact that the higher leaders of Junák were aware of its contents is evident from the view that the scout association was considered a preliminary organization of world Freemasonry, with an effort to guide it in that spirit.

Štekl's work begins with two quotes:

"The joy of the whole earth is Mount Zion, the heights of Zaphon; the city of the Great King."
Psalm 48:2.

"Jerusalem, Jerusalem, you who kill the prophets and stone those sent to you, how often I have longed to gather your children together, as a hen gathers her chicks under her wings, and you were not willing!"
Matthew 23:37.

The following text starts with Štekl's conjecture that the founder of the scout movement, Baden-Powell, whose real name was Robert Stephenson Smyth (a Freemason according to the information from the magazine of the Spanish Grand Orient), was "imbued with the two-hundred-year-old Masonic idea" when he developed youth education aimed at Wisdom, Strength, and Beauty.

Štekl celebrates Baden-Powell as the founder of a new youth religion that teaches "a return to the mysteries of ancient man." He states, "The mysteries of ancient Greece, with all the elements of Egyptian, Chaldean, and Jewish mysteries, permeate the entire scout's life." Štekl believes that the symbolism of the Scout Association's external insignia can be traced back to prehistoric times. For example, he mentions the staff, which is half an axe and half a hammer, given to specially honored scouts, stating, "Even the Chaldean god Ramman had an axe in his right hand alongside the moon and the sun in his symbols."

Overall, in his work, Štekl develops a unique symbolism of celestial bodies:

"The sun, moon, and stars were the cult of perhaps the first human being who had the ability to observe and reason. They were the cult of the ancient Chinese, Mexicans, Assyrians, Egyptians, and are still the cult of scouting, both in nature and in life. In life, our sun was the ideal of Freemasonry, President Liberator Dr. T. G. Masaryk, the protector and great supporter of the Scout Association. Our moon is the current President of the Republic, Dr. Edvard Beneš, the former mayor of the Scout Association until his election as president, and since then, the honorary mayor of the Association. He is now our second protector as the second president of the republic. Among the stars, we have the late Professor Dr. Karel Weigner, who wrote the book 'The Meaning of Scouting in Physical Education,' as our guiding pole star."

Again, the name of Grand Master of the National Grand Lodge of Czechoslovakia, Professor Karel Weigner, who passed away in November 1937, is mentioned here. He was previously referred to by Polak Filochowski in his work on Czechoslovakia as a leading Freemason, and we have encountered him again in the presidency of Sokol. He is now celebrated as the "pole star" of the Junák organization. The connection is evident. In civilian life, Weigner was regarded as an honest and quiet scholar. He was a professor of anatomy at the Czech University in Prague, dean of the Medical Faculty in 1922-23 and 1931-32, rector in 1936-37, and a member of various examination boards for secondary schools and teacher candidates. He served as the general secretary of the Czech Academy of Sciences, a member of the Czech National Council, and various other cultural and scientific organizations. He was an honorary citizen of many Moravian towns and an officer of the French Legion of Honor.

Given this open acknowledgment by Junák of its protectors, it is not surprising to find further Masonic divisions within the Czech Scout movement. The Scout oath does not expire when a Scout outgrows the youth groups; "the taking of the oath incorporates the Scout into the worldwide brotherhood" and binds them for life. "The Scout card opens the doors of all lodges of the worldwide Brotherhood, guaranteeing us friendly advice and joyful assistance." Junáci should be aware that "their hands and hearts are linked in a chain with the Scout brotherhood, encompassing the whole world."

Only through "tests" can a newly accepted member reach the rank of a leader, a "master." "Just as Adoniram excelled in the knowledge of all the works necessary for the construction of the Temple, so the Scout acquaints himself with all manual skills that are accessible to him and proves through tests what he has accomplished according to his assigned duties." Admission of a newcomer is not primarily determined by mere registration, but by a "probationary period," especially if they are not the son of a Scout. Only after three months, the newcomer must pass the test for novices before being admitted to the Scout oath.

After a minimum of six more months, a Scout can take the "second-class test" and when they reach the age of 14, they can undergo the third test of contemplation, known as the "three eagle feathers test." In constant imitation of lodge customs, the Scout's forest is considered a "cell of contemplation" where they must spend a day and night in solitude. With exemplary conduct, a Scout can subsequently be admitted to the "first-class test." This is how Štekl describes the five levels of Czech Freemasonry for youth, whose degrees are explicitly called "wage increases."

The admission ceremony is described by the author in a very characteristic manner:
"The day and the ritual of the oath remain indelibly engraved in the soul of every Scout. After the leader's address, the boy recites the words of the oath with his left hand placed on a book opened to the text of the oath. For us,

Czechoslovak Scouts, that book is Svojsík's Foundations of Scouting.

The Scout raises his right hand as if taking an oath, with three middle fingers extended while the thumb covers the nail of the little finger, which is also bent.

The three extended fingers represent the three clauses of the Scout oath, and the thumb covering the little finger symbolizes the duty of the stronger to protect the weaker."

This hand position accompanies the Scout throughout their life during the Scout salute, serving as a daily reminder of their oath. In the Scout uniform, they raise their hand to their hat in the same manner.

By simply raising their fingers in the prescribed position to their shoulder, the Scout gives the "secret salute," which is a means of communication between brothers who do not know each other.

The sealed oath is accompanied by the leader's handshake with the left hand, using the initiated grip to signify the union of heart and mind, and with a solemn formula, the newcomer is declared a member of the worldwide brotherhood of Scouts.

The exchange of the left handshake remains a secret touch among Scouts for their entire lives, replacing the common right handshake.

Special requirements were placed on leaders of Scout groups. "Calm and composed like a cornerstone, open like a set square, let the soul of the future Scout officer be." Only they may kindle the campfire in the evening or designate a brother to kindle the "sacred fire," greet the rising sun, and initiate the work of the Scout group, "whether in nature by the campfire or in clubrooms, which in many ways resemble the lodges of medieval stonemasons." Work itself becomes a genuine Scout act only if it is not disrupted by anyone "unauthorized," including the parents of the Scouts.

Let us not overlook the extent to which the Junáci truly lived in this ideology. It is essentially insignificant whether this was only the planning of their leaders' circle. What matters is the thinking that speaks through such ideas and at the very least testifies to the goals set by the leadership of the Czech Scout movement in 1938[iv].

Štekl was not some insignificant outsider in Junák; in fact, in 1938, when he wrote his document, he held the position of regional leader of the Scouts in Moravia. If we take that year as a basis for examining the individuals in leadership positions within Junák, we find that Jan Šváb, a Freemason, served as the regional leader in Silesia, and Hrabar, also a Freemason, was the regional leader of Junák in the eastern region of Czechoslovakia, specifically Subcarpathian Rus.

Similar patterns were observed at the district level. For example, the Freemason engineer Dr. Bohumil Glos served as the district leader of Junák in Olomouc. Above all, the central leadership of the Scouts was permeated with Freemasons. Dr. I. Charvát, a Freemason and university professor in Prague, held the position of chairman of the central council of Junák, among other Freemasons in leadership positions. Therefore, Štekl, a Freemason and Scout leader, was not an exception but rather a very typical and regular case.

The responsibility that adults assume before the history of their nation is clear and unquestionable. The inevitable end of the

Czech youth movement, whose ideals culminated in the concluding words of one of its regional leaders, was a consequence of prevailing circumstances. The leader's words address the Freemasons as "masters of the present age" and implore them to assist in building a new temple of peace, humanity, and love, which was the ultimate goal of their work. The reference is made to the ruins of Solomon's Temple and the trembling foundations of contemporary culture, highlighting the imminent collapse. The Scouts expressed their aspiration to be the "rough stones" in the construction of this new temple, guided by the Great Royal Art.

Semitic Czechoslovaks

In 1868, the writer Sir John Retcliffe (a pseudonym used by the then editor of the newspaper "Preussische Kreuzzeitung," (Prussian Cross Newspaper) Hermann Gödsche) wrote his novel "Biarritz." In this novel, he describes a secret nocturnal meeting of world Jewry, where they discussed the complete dissolution and planned destruction of Aryan nations. Due to its sensational content, this novel was reprinted after the World War, as it closely aligned with "The Protocols of the Elders of Zion," which were only revealed at that time. This suggests that Retcliffe already knew about the Jewish plans.

Retcliffe chose the Prague Jewish cemetery as the meeting place for the world's Jewish conspirators, demonstrating his knowledge of Jewish tradition. This Jewish cemetery, which still exists today, is located in the heart of the former Prague Ghetto, surrounding the Jewish Town Hall and the Old-New Synagogue, the oldest synagogue in Europe after the Amsterdam temple. The cemetery is a unique curiosity in the world and, according to Jewish tradition, served as a burial ground for a full 1200 years. Among the jumbled collection of stone monuments and gravestones, we find tombstones of great emancipators of Prague Jewry under Emperor Rudolf II, such as Mordechai Meisel, Rabbi David Gans, and, above all, the renowned Rabbi Jehuda Löw.

Rabbi Löw is associated with the old Jewish legend that as a magician of secret arts and a Kabbalist, he created the Golem. According to the Talmud, the word "Golem" means an embryonic, unformed substance, and according to the legend, the Golem was a human-like figure created by Rabbi Löw around 1580 from clay. It came to life when a "shem," the name of God, written on a piece of paper, was placed in its mouth. This Golem was believed to have the power to protect Jewry and destroy non-Jews (gója). However, when it became unpredictable and uncontrollable, the Rabbi extinguished the spark of life in his creation, causing the figure to collapse into a pile of clay, which, according to legend, still lies in the attic of the Prague Synagogue.

The Prague legend of the Golem has been repeatedly depicted in literature, especially Jewish literature. The previously mentioned Polák Filochowski, during his stay in Prague, tried in vain to uncover its true meaning. All the people he asked, including Jews, gave him only vague answers, and even the Jewish guide at the cemetery repeatedly said to him, "That is interesting, isn't it?" Filochowski, at the end of his account, wonders what happened to the Golem:

"Did it grow into the wall? Did it seek another embodiment? Is it perhaps waiting for its moment in that beautiful city where the liberated ghetto, uncertain about its newfound freedom, now governs in lodges of the alliance of democracy with communism, saturated bourgeoisie with insatiable revolution, peace with war?" (Filochowski, p. 276.)

And the Pole found his confirmation when he arrived at Barrandov, where a film about the Golem was being shot, financed by a Czech-Jewish-French company, and touted in the press as a pinnacle film and "a jewel of European film production" in advance.

Certainly, the clearest interpretation of the Golem myth was recently presented to us by Karl Hans Strobl in his charming book about Prague, published by Luser, Vienna-Leipzig in 1940, where he writes:

"However, today we can undoubtedly look into the Golem's innermost being with open eyes. Isn't it written in Jewish books: God did not want animals to serve His chosen people, and that is why He supposedly created the rest of us, the unbelievers, the impure, to obey the Jews as animals in human form? And wouldn't the Golem then be a symbol of those nations that are only called to life when they have to serve the Jews and only as far as their benefit requires? And if they rebel against Jewish orders and become disobedient, do they not sink back into an amorphous state? According to the ghetto's opinion, the Jewish spirit would be the shaping and animating power that awakens the sluggish mass of Gentiles at will to serve and, when it begins to gain consciousness, casts it back into the depths of matter. Could this be the ultimate meaning of the Golem legend and the ultimate mystery of the Jewish city of Prague?"

The decay of Aryan nations and the conquest of the most important positions in European state and national life by Jewry is inconceivable without international Freemasonry. It can even be said that lodges have almost paved the way for Jewish emancipation and, in recent decades, have become the political base of world Jewry[v].

For too long, Freemasonry has managed to hide this collusion and portray its exposure as baseless malice that cannot be politically or scientifically proven. However, since Freemasonry has been banned in Germany and its archives have been opened for public use, an irrefutable scientific proof has been presented that the international lodge system was nothing more than an accomplice to Jewish plans for world domination.

This scientific evidence has been particularly successful in the work of Dr. F. A. Six, a university professor in Berlin, in his treatise on Freemasonry and Jewish emancipation ("Freimaurerei und Judenemanzipation"), published in 1938 by Hanseatische Verlagsanstalt. Six builds his arguments based on the Magna Charta of the lodge system and the manifesto of Freemasons from 1723. He analyzes the role of Judaism in English Freemasonry and its infiltration into German lodges, through which it gained access to public life. He convincingly demonstrates how the ideologies of humanity and tolerance are nothing more than gateways for world Jewry into the spheres of European states and nations.

These clear scientific findings also offer a new perspective on the recent decades of Czech history and the country's internal national development. The ancient Prague Golem has risen from the dead in a new form, infiltrating the fabric of public life through the lodges and successfully destabilizing the Czech nation to the point where it was unable to address the Jewish question on its own terms.

It was precisely Czech Freemasonry of the Scottish Rite that enabled the development of Jewry in this region, which was influenced by Czechoslovak ideals. The generous acceptance of Jews into these purely Czech lodges stamped the Semitic brethren with a mark of absolute political reliability. On the other hand, the Jewish influence in the "Lessing u tří kruhů" Grand Lodge, which only prepared the first three degrees, was relatively stronger but of far lesser significance for the post-war Czech destiny.

Filochowski states that Jewish members constituted 40% of the National Lodge of Czechoslovakia. We have no reason to doubt this ratio today. The basis for this Jewish penetration lies in the fundamental concept of the "Národ" Lodge and the position of resistance circles during the World War concerning the Jewish question in general. From the study by František Sís, the founder of Czech Freemasonry, we can learn enlightening details about this. Sís informs us how the "Národ" Lodge addressed the Jewish question and how it dealt with it in one of its early meetings. Although the lodge did not yet have any Jews among its ranks at that time, any idea of anti-Semitism was "naturally" excluded from the outset. The writer continues:

"The Jewish question was presented for discussion as a serious national and religious problem, about which Henry Wickham Steed expressed in his book "The Habsburg Monarchy" published in 1915 that 'no man, be it a writer, politician, or diplomat, can be considered mature unless he sincerely attempts to understand the Jewish problem.' Regarding the religious aspect, all the speakers held a unified opinion that from a philosophical perspective, we have an equal objective relationship to every religion and every church, which is the relationship of religious tolerance, engraved among the doctrines of the first Masonic constitution."

By the "first constitution of Freemasons," Sís meant what Berlin university professor Dr. Six referred to as the "manifesto of Freemasons," namely the constitutional book by London Presbyterian preacher James Anderson (The Constitutions of The Free-Masons. Containing The History, Charges, Regulations, etc. of that most Ancient and Right Worshipful Fraternity. For the Use of the Lodges. London 1723), which, in its main part, the "Old Charges," established the ideological and moral foundations of Freemasonry.

These culminated—in rejecting "accidental" religion, nationality, and social status—in the doctrine of a purely rational world religion in which "all people agree" and thereby advocated for the fundamental equality of Jews. With these few sentences by Sís alone, the significant Czech voice joined the other pieces of evidence as a pivotal witness to the findings of Six's research.

The outcome of the meetings described by Sís was that the "Národ" Lodge took the position that Jews in Czechoslovakia are not a national or racial unit but rather a "product of a thousand years of spiritual and moral development." Their creative development was influenced by both the religious aspect, the idea of Jewish religion as such, and, secondly, by the national component, that is, Czech history, Czech culture, and Czech customs. "Jewish life was reflected in these two internal forces that interpenetrated and merged with each other."

During the discussion of this national aspect of Judaism in Czechoslovakia, the "Národ" Lodge, as the precursor of the future Czechoslovak Grand Lodge, arrived at ideas and principles that had a decisive impact on the future development of the Czech national stance towards Judaism. Sís expressed it in these words: "If their result showed that the souls of Jews are permeated by Czech national thinking, Czech moral sentiment, the spirit of Czech culture, and Czech historical traditions, if their thinking is demonstrated through actions in civic life, then they are the product of Czech national development and a part of the Czech nation's organism."

This pronounced the fundamental equality of Judaism in Czechoslovakia with lasting national consequences! If the "Old Charges" of the manifesto of Freemasons from 1723 formed the basis of the emancipation of world Jewry, then this resolution by the first Czech Lodge in 1919 was the recognition and implementation of the "Old Charges" in favor of Jewry at the birth of the Czechoslovak Republic.

Practically, this resolution meant for the "Národ" Lodge, the Czechoslovak Grand Lodge, and the Supreme Council for Czechoslovakia a willingness to accept all Jews, and it served as the starting point for the emerging Jewish penetration into the Czech Grand Lodge. Although the resolution outwardly contained restrictions in the sense that Jewish seekers should not have been "national cowards" during the World War and that they "bravely fulfilled their national duties" during times of danger, this limitation was essentially insignificant. It corresponded to the general and uniform demand of the Czech Lodges of the Scottish Rite, a demand that was also placed on all Czech brethren.

Further considerations in Sís's study on the Jewish question extensively deal with the distinction between "German Jews" and "Czech Jews". During the World War, a large number of them were in the Czech-Austrian activist camp. Sís himself attributes only historical significance to this question in terms of the principles of Freemasonry. For us, the essential point is the acknowledgement that Jews were also part of the Czech revolutionary camp, and that even the Mafia had Jews among its collaborators.

The limited scope of this explanation does not allow us to describe all the details of how Jews began to infiltrate all aspects of Czech life since the founding of the republic. It happened step by step as the influence of Freemasons expanded, in line with the principle set by the "Národ" Lodge, which called for not only complete equality but also national inclusion. Thus, the term "Semitic Czechoslovaks" and "our Jews" emerged, which were approved by the state and society.

In one aspect, however, this development over the past two decades can be briefly indicated as an example. Once again, we turn to the Sokol organization, the largest and richest Czech organization with a long tradition. Its infiltration by Freemasonry has already been described, and it was suggested that the ideological decomposition by Freemasonry was accompanied by Jewish infiltration from within the membership. Indeed, Jews, often in large numbers, were accepted into local Sokol units, gradually infiltrating and eventually occupying leadership positions. A Jew named Heller even became a member of the Czech Sokol community's presidency. At the same time, Jews were particularly welcomed guests at various Sokol events. In contrast, Sokol gyms were closed for lectures of a national and antisemitic nature.

In the southern regions of Bohemia, it even happened that a Jew became the founder of a Sokol unit. Jewish individuals, alongside Freemasons, also wrote "enlightening" and politically subversive articles in the journals of the Sokol organization. The highest-ranking Sokol officials consistently exposed themselves professionally to their Jewish colleagues and others. Recently, the Protectorate press has published such a significant number of documents of this nature that it is unnecessary to provide more examples here. The gradual Jewish infiltration and disintegration of the largest Czech sports organization was completed over two decades.

We will mention one characteristic excerpt from the mentioned press discussion. The Prague newspaper "Národní politika" (National Politics) published an article on November 2, 1941, titled "Jewish Infestation of Sokol," which also included a non-anonymous contribution from readers. The author, a director by profession and an active member of a rural Sokol unit, stated, among other things: "As a Sokol member, I clashed with the then district education officer in 1928 during a Sokol meeting in Pardubice because I did not approve of Jewish representation in this unit. I received the response: 'These are the Jews who are national Jews'... Unfortunately, I was left alone in my fight against the Jews." Thus, the "Old Charges" from the constitutional book of 1723, adopted by Czech Freemasons at the birth of the republic, already bore their ideological and political fruits, even in the rural units of the Sokol organization, after a decade of the existence of the Czech state. This significant Czech national organization confirmed Jews in their affiliation to the Czech nation.

This development reached its visible peak during the X All-Sokol Slet (Sokol Gathering) in Prague in the summer of 1938. A strong delegation from the Maccabi Association in Palestine also attended the event and brought greetings from the Jewish community in the Promised Land. The "Jewish Correspondence" from Belgrade enthusiastically described this event, stating: "A numerous delegation from the Maccabi Association in Palestine, an organization similar to Sokol, also participated in this grand celebration of the entire Sokol movement. Representatives of Jewry from Palestine conveyed greetings to the Czechoslovak nation and Sokol, for which Jewry has a particularly warm sympathy." This acknowledgment from the world Jewish community to Sokol signifies nothing other than a conscious gratitude for Sokol's willingness to assist in the political and power-driven development of Jewry in the Czech Republic. In the same spirit, the Prague Jewish magazine "Selbstwehr" published an article expressing greetings to the Sokol celebrations, highlighting the exceptionally friendly stance of the Czech Sokol leadership towards a similar Jewish revival movement. The article also mentioned that many representatives of Czech Sokol were excellent friends of the Maccabi Association, and that many Sokol members worked as teachers in Maccabi.

Then, several months later, after President Beneš fled from his political responsibility to his homeland, the Second Republic, under pressure from Czech nationalist circles, hesitantly and ultimately without significant results attempted to address the Jewish question. Voices of reasonable people emerged, calling for the cleansing of their great Czech sports organization from Jewish influences. Suddenly, the press reported that the leadership of the Czech Sokol Community requested a resolution to the Jewish question during a meeting on Sunday, October 23, 1938. Considering the Jewish-Freemasonic subjugation of Sokol, it was clear that this could only be a maneuver to deceive the public. A closer examination makes it evident. During the aforementioned meeting of the Czech Sokol Community, with the participation of 97 representatives from all regions, the following resolution was adopted:

"The Jewish question should be addressed on the basis of nationality and social status, so that immigrants who arrived after 1914 return to their original homelands, and among the remaining Jews, those who declared Czechoslovak nationality in 1930 should gradually integrate into the social stratification of our nation in proportion to their numbers, while others should go to the countries they voluntarily affiliated with in 1930" according to "Pražský list" (Prague News) on October 24, 1938.

Once again, these principles served as godparents, as explained by Sis in the early days of the "Národ" lodge, as a manifestation of the "Old Duties." The Sokol resolution of October 23, 1938, practically meant that the Sokol leadership and representatives of the regions decided to accept the long-established, nation-building Jewry of the Czechoslovak Republic as part of the nation. The "National Jews" were simply Czechoslovaks!

One might have expected that after the establishment of the Protectorate, a corresponding cleansing of Czech physical education life would take place. It also seemed that way on the surface, with Sokol determined to carry out a serious purge. The Sokol leadership issued an official statement stating that as of August 26, 1939, all Jews would cease to be members of Sokol, they would be removed from membership lists, and they would be notified of their expulsion by registered letter. However, even this was just a superficial maneuver carried out under pressure from nationalist circles. For example, only one right-wing Czech group submitted six sharp petitions to the Sokol leadership, demanding a radical purge to finally take place.

It is also an open secret what hidden thoughts and plans the Sokol leadership had during their plenary session in Prague when discussing the Aryanization decree. There were tumultuous scenes at the general assembly of the Budějovice Sokol when the right-wing faction presented their demands. The practical results of the so-called "Jewish purge" were also evident: Exposed Jewish Sokol members were outwardly marginalized, but everything else remained unchanged.

However, there was one aspect in which the external stance of Czech Sokol had changed in recent years: the leadership avoided openly expressing a friendly position towards Jews to avoid compromising themselves before the Germans. Yet, as observed in such cases, secret messages were passed on to Sokol brothers abroad, allowing them to proclaim the old slogans more loudly. The Sokols in America initiated this cycle in their bulletin by affirming their commitment to proven principles of democracy, equality, and brotherhood (Sokol americký, Volume IX, Issue 11, April 15, 1939), with no doubt about their fragile worldview. However, the Czech Sokol in Paris achieved a significant piece regarding the Jewish question. In the January issue of their bulletin (Volume III, Issue 1, January 20, 1940), they extensively elaborated on Sokol's stance on the Jewish question. It is interesting to note that they admitted right at the beginning of the article "Us and the Jews" that even their unity had not been spared inappropriate remarks that the Paris Sokol would soon become a Jewish platform. The main part of their further reasoning is cited verbatim for its "principle."

"To propagate anti-Semitism as a principle, that is, to identify with Hitlerism, Sokol rejects - just like any decent person. According to its statutes, anyone can become a member of Sokol, regardless of their religion or political beliefs. Jews who have Czech or Slovak as their mother tongue, or those who not only speak but also think and share Czech or Slovak values, and who fulfill other requirements (moral integrity, humanity, national loyalty) can be full members of Sokol, just like Jews who are considered equal and respected members of our nation, having been raised, feeling, and thinking in Czech or Slovak, and many of whom no longer identify as Jews even in a religious sense because they have also left the Church, proving themselves to be just individuals and deserving patriots. Sokol does not divide humanity based on race, color, appearance, or class, but based on character, intellectual capacity, and moral values. Therefore, our attitude towards good Czechoslovak Jews, proven patriots, can only be positive and good."

Even this latest official statement by Sokol regarding the Jewish question seems to be in line with the resolution of the "Národ" lodge on implementing the manifesto of the Freemasons, in which Sokol Mayor Dr. Scheiner also participated. With this latest acknowledgment of Sokol towards "good Czechoslovak Jews," it is certain that Sokol has fully fulfilled the "Old Obligations" manifesto.

The old Jewish legend of the Golem, "this true Prague native," as Strobl calls him, took on a new meaning in the twenty years of the independent Czech state. This is evidenced by the history of Sokol, as well as the relevant development in all other areas of Czech national life. In Czech lodges, the Golem arose and, in the spirit of its creator, Rabbi Löw, enslaved the Czech gentiles, bewildering their minds and rendering them incapable of freeing themselves from Jewish domination.

Is there a more fitting testimony for this than the fact that the creator of the monument to Hus on the Old Town Square erected a monument to the creator of the Golem, Rabbi Löw, on the corner of a building in the former Prague Jewish Town?

Golem above the Castle

The Czech Freemasonry, later known as the National Grand Lodge of Czechoslovakia, originated from the activities of the Czech resistance during World War I. The members of Maffie and the "Národ" group were its founders and initial spokespersons, as they were almost entirely accepted as seekers. International Freemasonry recognized the activities of Maffie as a testament to the principles of Freemasonry.

In this sense, lodges were established, at least in the early years, with strict political exclusivity. According to the "Old Obligations," the fundamental law of Freemasonry, Judaism was also embraced as long as it appeared Czech.

The deliberate penetration of public life in the Czech Republic stems from the initial plan for the expansion of the "Národ" lodge in 1919, aiming to establish new lodges based on professions to quickly create centers of Masonic work in various areas of life. Once these lodges were approved by the National Grand Lodge of Italy during Christmas of 1919, the conquest of decisive key positions in all fields began, culminating triumphantly after twenty years of effort. Undoubtedly, the most remarkable achievement was the successful infiltration of Sokol, the main Czech physical education organization, and the youth movement, both of which, after being personally intertwined and adopting the spiritual heritage of Freemasonry, were alienated from their true national tasks. After being ideologically poisoned and corroded, they were lost as potential seeds of true Czech nationalism. Simultaneously, the conquest of the economic sphere occurred, as it plunged into the boundless current of liberal economic ideas and methods. The cultural realm was no exception, with Czech artistic life becoming a stronghold of unbridled culturally Bolshevik tendencies. The same can be said for the political sector.

It would be tempting to trace this line of development, which we observed in Sokol and Junák, in all other areas of Czech national life. However, delving into the details would take us too far within the given framework. President Masaryk, as Štekl's pamphlet for the Masonic youth of the scouting movement put it, shone as a radiant sun for all other fields of national life.

On the contentious question of whether Masaryk also belonged to Freemasonry, we will only briefly mention it here. Many anti-Masonic publications claimed that he was a member of the lodge, and this view gains some credibility from the fact that the office of the Czechoslovak National Council during the World War was located in the building of the "Grand Orient de France" at 16 rue Cadet in Paris. In the International Lexicon of Freemasons, published by Freemasons themselves, it is denied that Masaryk was a lodge member, but it is emphasized that his way of thinking completely coincided with the way of thinking of Freemasons. Additionally, certain French lodges organized memorial ceremonies after Masaryk's death, in which he was referred to as a Freemason. The late Grand Commander of the "Supreme Council for Czechoslovakia," Prof. Alfons Mucha, himself stated, based on several conversations with Masaryk, that Masaryk did not join Freemasonry out of respect for several old friends from his time studying in Vienna, who were already members of Freemasonry or humanitarian associations established to conceal it, and they urged Masaryk to join, but he refused. The complete alignment of his way of thinking with the ideology of Freemasons also led to him being called the "Freemason without an apron" in Masonic circles.

For the development of Masonic power and influence in Czechoslovakia, the question of whether Masaryk formally belonged to Freemasons had little significance. However, shortly after his return home, Beneš joined the "J. A. Komenský" lodge in Prague and became a 33rd-degree Freemason. In him, the influence of Freemasonry in the Prague government was present from the beginning of the republic until its end.

Under his long-standing patronage as the Czech Minister of Foreign Affairs, his ministry was the first to be infiltrated and penetrated by Freemasons. This was particularly true since several Freemasons were active in the Ministry of Foreign Affairs alongside Beneš. Consider, for example, the case of Dr. Kamil Krofta, whom we saw as part of the first group of seeking brothers in the "Národ" lodge in 1919 and who served in the Czech Ministry of Foreign Affairs from 1920. In the same year, he was appointed as the ambassador to the Vatican. In 1921, he became the ambassador to Vienna, and in 1925, he served in Berlin. From 1927, he held positions in the Ministry of

Foreign Affairs as Beneš's deputy and head of department. When Beneš was elected president of the republic in 1935, Krofta became his successor as the Minister of Foreign Affairs. Consequently, the ministry remained consciously under the control of the high-degree Scottish Freemasonry, which did not change even after Krofta's resignation. Another close collaborator of Beneš, the late Jiří Sedmík, a councilor in the Ministry of Foreign Affairs, was a high-degree Freemason and served as the Grand Secretary for Foreign Affairs in the National Grand Lodge in 1937.

The same picture emerged within Czech diplomacy abroad. Jan Masaryk, the son of the first president of the republic, was a Freemason according to a letter from Alfons Mucha, a 33rd-degree Freemason. He served as chargé d'affaires of the Czechoslovak Embassy in Washington from 1919 to 1920, was active in the Prague Ministry of Foreign Affairs, and became the ambassador to London in 1925. Let's also consider the founding member of the "Národ" lodge, Dr. Vladimír Slavík, who was elevated to the rank of Master Freemason by the National Grand Lodge of Italy. He served as a secretary for the Czechoslovak Peace Delegation and was a member of the political section of the Secretariat of the League of Nations until 1928. From 1928 to 1931, he headed a department in the Czech Ministry of Foreign Affairs and then went to Brussels as an ambassador. Today, his portrait is exhibited in the European Museum of Emigration in London.

The deliberate personnel policy of Freemasonry within his ministry is demonstrated by the case of the lodge's founder, František Sís, who rejected a diplomatic career offered to him only on the instruction of his leader in the National Democratic Party, Kramář. Sís himself stated in a personal note, which is among the materials of this conspiracy, the following:

"The decision to stay in Paris resulted from an offer made to me by the Minister of Foreign Affairs, Dr. Ed. Beneš, in March 1920, to go as an ambassador of the Czechoslovak Republic to Sofia. President Masaryk, during a reception at the Castle, engaged me in conversation and advised me to go to Sofia. 'If I were young,' he said to me, 'the Balkans would be where I would like to be active as a diplomat.' Some of my friends recommended it to me, especially the current ambassador to Brussels, Dr. V. I. Slavík, who discussed it with Dr. Beneš.

Everything was already settled, and some newspapers announced my appointment."

If the mentioned names alone testify to the instructive and fateful connection between Freemasonry and foreign policy in Czechoslovakia, it can also be supported by documents. An interesting example is the report from the Grand Lodge of Bulgaria dated October 24, 1928, regarding tensions with the Grand Lodge of Yugoslavia over the Macedonian question. This report was sent to the National Grand Lodge of Czechoslovakia in a letter dated January 29, 1929, addressed to the Deputy Grand Secretary for Foreign Relations, Kamil Krofta, requesting his foreign policy stance considering the interests of Czech Freemasons. The National Grand Lodge of Czechoslovakia also sought the relevant approval from the Minister of Foreign Affairs, Freemason Dr. Beneš.

This infiltration of the Czech state apparatus by Freemasons was mirrored in parliamentary life in Prague. Even a glimpse at the revolutionary National Assembly from 1918 to 1919, whose members largely played leading roles as Czech politicians in the following two decades, provides a clear picture. Let's consider the club of National Democrats, who emerged from the merger of the State Law Democracy (former Young Czechs), Old Czechs, and a faction of the Realistic Party.

We can see the following image of how the first parliamentary body of the republic was permeated by Freemasons: Among the 46 members of the National Democratic Club, there were, for example, František Sís, the founder of Czech Freemasonry; J. S. Machar, the first Grand Master of the Czech Grand Lodge; Dr. J. Scheiner, the deputy Grand Master and Mayor of Sokol; 33rd-degree Freemasons Dr. Bohumil Němec, Dr. Přemysl Šámal, and Dr. Alois Rašín; Viktor Dyk, a recorder and Knight of the Rosy Cross, an 18th-degree Freemason; Jaroslav Kvapil, the author of the resistance manifesto "To the Czech Delegation in the Imperial Council" dated May 17, 1917, and Chief of Drama at the National Theater, among others.

Although the infiltration of Freemasons in the parliamentary club of National Democrats was the most conspicuous case, it was not the only one in the revolutionary parliament. In subsequent electoral periods, this foreign influence expanded significantly among other parties, particularly among the National

Socialists aligned with Beneš. However, the example mentioned above also demonstrates how Freemasonry influenced political parties. Internal conflicts within parties and subsequent upheavals can be traced back to this influence. Let us return to the example of the National Democrats, whose leader, Dr. Kramář, was one of the few Czech politicians who resisted the influence of Freemasonry. As a result, his group declined both in numbers and significance. Especially when they joined forces with the Silver League, Freemasons continuously departed from the group and sometimes even turned into fierce opponents. Prior to the parliamentary elections of 1935, 35 Freemason intellectuals left Kramář's "National Unity" and openly switched to the National Socialists, who were loyal to Beneš.

Shortly thereafter, Grand Master Dr. Karel Weigner issued a proclamation to the Freemasons of Pilsen opposing the system of party-affiliated candidate lists. However, Weigner, then the rector of the Czech University in Prague, did not want to expose himself further in this Freemasonic action and instead put forward Professor Dr. Václav Hora, the dean of the Faculty of Law at that time, who announced the establishment of the "Society of

Free Elections." Hora had previously served as the Master of the Lodge "Národ" and later rose to the rank of Grand Master of the National Grand Lodge of Czechoslovakia. However, his Freemason-inspired attempt ended in failure.

In the parliament, hardly any politicians dared to defend against Masonic raids. The secretary of the Agrarian Party, Deputy Ing. Žilka, issued a lone warning during the session of the parliament on October 30, 1936, stating that it should not be tolerated when groups and lodges are formed outside of legal positions, which only harms public opinion and the interests of the state. Besides the government and parliament, there was another place of political authority and influential power in the Czech Republic, which, despite its outwardly low visibility, referred to itself as the "third instance" and was celebrated: the Czech National Council. It was founded in 1900 and became an important working institution in national politics even before the World War. During the underground activities of domestic resistance from 1914 to 1918, it proved to be effective and developed with its newly established branches in Moravia, Silesia, Slovakia, and Carpathian Ruthenia into the highest leading and directing body of all Czech cultural and national work, which it remains to this day. Thus, it represented a third political position for Freemasonry alongside the government, administration, parliament, and political parties, which needed to be conquered. Again, immediately after the revolution, the same well-known names emerge, adorned with the glory of national pioneers.

In 1919, Freemason Dr. Přemysl Šámal already occupied the position of Deputy President of the Czech National Council, with high-degree Freemasons J. S. Machar, Dr. J. Scheinen, and Dr. Vávro Srobár standing by his side as members of the council's committee, along with several brethren from lower-degree lodges. Freemason Dr. Rudolf Pilát also sat on the control commission. Soon, Weigner, Kapras, and Němec followed and joined the committee, to the extent that the influence of Freemasons became significant. Freemason Dr. Němec, a 33rd-degree Freemason, was eventually appointed to the chairman's seat of the Czech National Council. When Němec considered it appropriate to somewhat step back after his unsuccessful presidential election campaign in December 1935, Freemason Dr. Jan Kapras, who had advanced to the 33rd degree and had become the Deputy Grand Commander of the

Supreme Council for Czechoslovakia, assumed the position of President of the National Council. Němec, however, remained in the committee, and Freemason Professor Dr. Josef Charvát joined him. Under the guise of university chairs and cultural-political work, the decisive central working body of the Czech Grand Lodge gathered there for twenty years, firmly holding onto this pinnacle position once it was attained.

These brief glimpses into the individual political spheres of the republic convincingly demonstrate how Freemasons systematically infiltrated all the crucial key positions in Czech state and national life. The methods used and implemented, which sometimes involved placing scarcely qualified lodge members in positions, are evidenced by the careers of that highest group of Freemasons who were elevated to the highest degrees of the Scottish Rite by the National Grand Lodge of Italy in Rome in 1919. The first Grand Master of the Czech Grand Lodge, J. S. Machar, became the General Inspector of the new Czechoslovak Army and was a candidate for the Minister of National Defense during the formation of the second Tusar cabinet. His deputy in the position of Deputy Grand Master, long-time Mayor of

Sokol Dr. Josef Schemer, took over the Czech Army after the revolution and became its first General Inspector, preceding Machar. We have already followed the life story of the first Grand Tyler of the Grand Lodge, František Síse.

Freemason Dr. Přemysl Šámal, a 33rd-degree Freemason, became the Chancellor of the President of the Republic in February 1919, and he held this position, not just as a ceremonial role, even after Masaryk's resignation under Beneš. Freemason Dr. Alois Rašín, also a 33rd-degree Freemason, became the Minister of Finance in the Kramář government after the establishment of the state, and he held the same position in the first Švehla government in October 1922. His anti-German chauvinism ("I don't negotiate with rebels!") severed all ties between the Czechs and Germans and immediately pushed the German Social Democrats into opposition after the revolution. The case of the assassination in which Rašín fell victim as a minister in January 1922 is still unresolved. Freemason Ing. Jan Dvořáček, a 32nd-degree Freemason, became the Minister of Trade in the second Švehla government. The list could go on as desired.

However, the most telling is the political journey of Freemason Dr. Bohumil Němec, a 33rd-degree Freemason. In his civilian life, he was a university professor of plant physiology and a long-time chairman of the Czech National Council. After the revolution, he joined the revolutionary National Assembly as a member of the National Democratic Club. After serving as a senator for some time, he left his party when it merged with the radical, Beneš-hostile Silver League. Since then, Němec was considered politically unaffiliated. It was in this context that he was called upon to run for the presidency of the republic in the autumn of 1935, marking the most significant situation for the Czechoslovak Republic since 1918.

During Masaryk's presidency, the Czech state adopted a consistent line, and the question of its continuation or interruption was being discussed. Freemason Dr. Edvard Beneš, a 33rd-degree Freemason of the Czech Grand Lodge, was being considered as Masaryk's successor, while Dr. Bohumil Němec, a 33rd-degree Freemason of the same lodge, was to be his rival candidate. Could they end up opposing each other? Unlike Beneš, Němec was not publicly known as a high-degree Freemason. And so, the game began!

On November 21, Masaryk informed Prime Minister Hodža in a hearing that he had an irrevocable intention to resign from the presidency and recommended Beneš as his successor. The Agrarians, as the strongest party, stated that Beneš was the vice-chairman of the Czech National Socialist Party, and therefore, a party member, which they found acceptable as long as he was a politically untainted personality. Suddenly, the name of the Chairman of the supra-party Czech National Council, Professor Dr. Bohumil Němec, emerged. Vraný, an exponent of the right-wing faction within the Agrarian Party, was the first to mention his name; it is unknown whose advice he followed. I will not go into further detail about the events, as Harry Klepetář, a Jewish journalist and author of the history of the Czechoslovak Republic, can provide a more comprehensive account.

"At the end of November, the crisis intensified extraordinarily. Hodža's efforts within his own party proved futile, and a fateful event was approaching: the contentious vote for the presidency of the republic, dividing the population between the right and the left. Nervousness among the leading statesmen grew day by day. Only Beneš remained calm." (Harry Klepetář, "Seit 1918 ... Eine Geschichte der Tschechoslowakischen Republik." (A History of the Czechoslovak Republic) Publisher Julia Kittla nást, Moravská Ostrava 1937, page 397).

Klepetář describes the behavior of the Agrarians as follows: "Beran, to the surprise of many people—including Prime Minister Hodža—approved all of Vraný's proposals, especially the proposal for the Agrarian Party to declare Prof. Němec as its official candidate. Now Masaryk intervened again. On December 2, 1935, he received Beran at Lány Castle. Nothing is known about their conversation... Since Beran did not advocate for Beneš' candidacy then or later, it can be assumed that he did not commit himself to Masaryk and that the President failed to win Beran over to his plans".

Němec played the role of a staunch candidate who focused the opposition against Beneš on himself.

"In the meantime, Hodža continued to strive for coalition unity on the issue of the presidency. He placed his hopes on Prof. Němec, expecting him not to accept the candidacy against Beneš and thus against Masaryk's wishes. When, on December 10, Němec agreed to his candidacy for the presidency, Hodža considered his efforts to have failed".

However, the situation became even more intricate and thus seemingly more truthful: "On Wednesday, December 11, Professor Němec stepped down as the chairman of the National Council in order, as he stated, to compete for the highest office in the state as a simple citizen and not as the leader of the supreme national organization. On Thursday, December 12, a crisis occurred in the National Council. During the session of the presidency, representatives of the Czech Social Democrats and National Socialists declared that Chairman Prof. Němec had violated the objectivity he was obliged to maintain by running as a candidate of the right-wing against President Masaryk and the majority of the nation, represented by Beneš. As a result, both socialist parties left the National Council. Representatives of the legionnaires joined them".

Now the game of high-degree Freemasons in the National Council began: "The presidency of the National Council, without these three groups, expressed the belief that the candidacy for the office of the President of the Republic is the right of every citizen and that there is no reason for Chairman Prof. Němec to resign. Prof. Němec took note of this, withdrew his resignation, and thus ran as the Chairman of the National Council in the presidential election".

At this point, Freemason Přemysl Šámal, a 33rd-degree Mason, took action. As the Chancellor of the President of the Republic, he accompanied the Prime Minister and the heads of both chambers to President Masaryk at Lány Castle on December 14. In a solemn act, he read aloud the President's declaration of resignation, which included the sentence: "I would like to tell you that I recommend Dr. Beneš as my successor. I have worked with him abroad and at home, and I know him".

Klepetář continues his narrative:

"On the same day that Masaryk resigned, the National Assembly was convened to elect a new President for December 18, 1935, in the Vladislav Hall. There were only four days left for the election campaign. The central organ of the Agrarian Party, 'Venkov,' wrote on December 15 that the Agrarians and the Craftsmen's Party support a non-political candidate and will propose the Chairman of the National Council, Dr. Bohumil Němec, for the position. They stated that Masaryk's recommendation of Beneš's candidacy was of a private nature".

Now the undermining of Prof. Němec's voter base began. Sokol and the Legionnaire Association were the first to address their appeal to the eligible parliamentary members, urging them to vote for Beneš. The advice of opposition groups seemed uncertain, but Němec did nothing to keep them united. He only sought someone to blame and believed he had found it in Hlinka's People's Party. Father Hlinka saw through this chess move and immediately responded with a statement that an insincere game was being played with the candidate of the so-called Civic Bloc, trying to shift the blame onto his party, the Slovak autonomists.

The mask fell! At midnight on December 17th, just a few hours before the election, the Czechoslovak Press Office issued the following statement:

"Prof. Bohumil Němec, Chairman of the Czechoslovak National Council, who was urged by political parties to run for the office of President of the Republic, set a condition that he would only maintain his candidacy if he became the candidate of the majority of Czechoslovak political parties. The indecisive behavior of one political party prompted Prof. Bohumil Němec to request that other political parties withdraw their support for his candidacy to ensure that national unity was not weakened".

Thus, the only opposing candidate withdrew just a few hours before the election. With only a few hours left until the electoral act, Beneš became the sole candidate for the presidential seat of the republic. On the morning of December 18, 1935, the election took place in the Vladislav Hall of Prague Castle: out of the 440 votes cast, 340 were for Beneš, 24 national democrats voted for Němec as an informal protest despite his withdrawal, and 76 ballot papers were blank, representing the Sudeten German Party, Czech fascists, and three unknown voters, likely Vraný and two other right-wing agrarians.

With the depiction of these events, Klepetář concludes his book. Shortly after its publication in 1937, it was banned. Why? Klepetář was, after all, a loyal supporter of the Masaryk-Beneš line. Did he perhaps describe the events too openly? Was this, the only comprehensive picture of the events surrounding the presidential election of 1935, meant to be erased?

In the state act of December 18, 1935, the disintegration of Czech nationalism reached its climax. The mafia, lodges like "Národ," the Grand National Lodge, and the Supreme Council found their culmination here after a decade of influence.

All contributed to the help in the saddle of the "moon" of Czech youth: Freemasonry, Judaism, the "brother without disguise" in Lány, the government, the parliament, political parties, the National Council, a clique of university professors, legionnaires, and Sokol.

Everything passed like ghosts! The old Prague Golem fell apart into a pile of clay. It will never rise again.

Illustrations from the Book

The accompanying illustrations have been scanned from the book, and each is accompanied by descriptive captions. Although the author provided translations and transcripts for some of the documents, not all of them were included. To translate the non-English and non-Czech text, I have sought the assistance of my colleagues. Due to the fading and illegibility of certain parts of the documents, our best efforts have been made to share their contents.

viz „Právo lidu" 20./4. 1919

941-O / 1

ÚVODNÍ SLOVO
k výroční zprávě vrch. vůdce, A. B. Svojsíka
jednohlasně schválené výborem i val. hromadou spolku
„Junák - český skaut" 28. listopadu
1918.

Prve než dospěji k podrobnému vylíčení práce naší v roce minulém, budiž mi dovoleno několik slov úvodem:
Po prvé dovoleno jest mi mluviti jako člověku opravdu svobodnému, na půdě svobodné k přátelům-skautům — občanům svobodné republiky československé. Po prvé možno tu mluviti otevřeně.
Radostno jest celkem vzpomínati *začátků* junáčení. Vůdcové našeho národa okamžitě postřehli, jaké kouzlo mocné stajeno jest ve výchově skautské, jak důležitým prostředkem pro vývoj a řekněme i nápravu povahy české mládeže může se junáčení státi. Vřele uvítal mé první články a plány dr. Kramář. Když nesl jsem základní dílo anglické »Scouting for Boys« Masarykovi, přivítal mne slovy: »Ach, dílo to již znám a budu skauting podporovati u nás, kdekoli mám vlivc. Návratu mládeže k přírodě těšili se upřímně Čáda, Drtina, Jirásek, Klofáč a jiní naší předáci jevili o hnutí nejen platonický zájem, ale i ruku činně k dílu přikládali.
Hůře bylo již s širším obecenstvem. Nescházelo výtek, neporozumění, útoků časopiseckých — i osobních — sesměšňování, ba i pohrůžek. Jádro však skautingu je tak zdravé, že vážnou prací pomalu sice, ale přece získávána půda a zjednávána důvěra i sympatie. *Byly tu však též rakouské úřady.* Dobře se pamatujeme na tu rakouskou byrokratickou úzkostlivost v ohledu výchovy mládeže — není tomu tak dávno, co jsme pod ní všichni strádali. My zaváděli jsme samosprávu mládeže, její schůzky v klubovnách, kroj, život ve volné přírodě a mnohé jiné, co hraničilo na paragrafy školských zákonů a disciplinárních řádů. Našli se i v úřadech jednotlivci, kteří dovedli rázem pochopit, co dobrého je ve skautingu. Ale rakouský systém žárlivě střežil každé hnutí mládeže, aby nevybočovalo z kolejí, do nichž je nutila hydra, jež v ohromném náporu právě chystala se ovládnouti celý svět. Nebylo tu jistě snadno postupovati, neboť stačil škrt perem, *aby celé slíbené hnutí* dostalo se tam, kde jsme je míti nechtěli, nebo aby již v zárodcích, jako »státu podezřelé a nebezpečné« bylo zaškrceno. Jaké

1

Revolutionary speech by Scout leader Svojsík on November 28, 1918.

Document text:

Introduction
to the Annual Report of the Chief Leader, A. B. Svojsik

unanimously approved by the Committee and General Assembly of the association "Junak-Czech Scout" on November 28, 1918.

Before I delve into a detailed account of our work in the past year, allow me a few words by way of introduction:

Firstly, let me speak as a truly free person, on the grounds of freedom, to my scout friends - the citizens of the free Czechoslovak Republic. Here, I can speak openly.

It is truly joyful to recall the beginnings of scouting. The leaders of our nation immediately recognized the powerful magic inherent in scout education, and how scouting could become an important tool for the development and even the correction of the Czech youth's character. Dr. Kramar warmly welcomed my first articles and plans. When I presented Masaryk with the foundational work of "Scouting for Boys" in English, he greeted me with the words: "Ah, I already know that work, and I will support scouting here wherever I have influence." Čada, Drtina, Jirasek, Klofac, and other leaders among us not only showed a platonic interest in the movement, but actively contributed to the cause.

It was more challenging with the wider public. There were criticisms, misunderstandings, attacks in the press - even personal ridicule and threats. However, the core of scouting is so strong that through serious work, slowly but surely, we have gained ground and earned trust and respect, overcoming the narrow-mindedness of the Austrian bureaucratic mentality when it comes to youth education - not long ago, we all suffered under it. We introduced youth self-governance, club meetings, uniforms, life in nature, and many other things that bordered on the restrictions of school laws and disciplinary regulations. There were individuals in the authorities who quickly understood the value of scouting. However, the Austrian system jealously guarded every youth movement, making sure it did not deviate from the mold it imposed, as it was preparing to dominate the world. We did not always progress smoothly, as a single stroke of the pen was enough to stifle the promising movement, labeling it as "suspicious and dangerous to the state." Thus, it was stifled in its infancy, against our will, because it was seen as a threat. (END) z Vídně v uniformách generálských neb s tituly baronskými a hraběcími.

Tu pak uprostřed nejhouževnatější práce a slibného rozkvětu našeho mladého ústředí přišla *válka* a s ní onen všeobecně známý *tlak shora*, který drtil a dusil vše, co jen česky dýchalo.

Skauti, kteří činně súčastnili se života veřejného, všech národních sbírek, byli při všem, kde jich bylo třeba, nemohli zůstati prostě stranou, když jiní, zejména i mnozí naši hrnuli se do služeb »ohrožené vlasti«. Zbývalo jedno z dvojího: Buď ustat v činnosti a začít za neznámých poměrů a neznámých okolností po letech znovu, nebo najít přijatelný modus vivendi, aby čest zůstala uchráněna a součinnost s ostatním světem »rakouským« aspoň zdánlivě zachována. Byla to domácí revoluce, tichý vzdor, který, když museli umlknouti i poslanci a noviny, provozoval kde kdo. V tom moři lidské bídy, které se světem rozlilo, výbor náš svorně a v souhlasu s našmi *ideami a skautskými sliby* rozhodl se pro činnost *humánní*. Byly to akce ve prospěch raněných invalidů, sirotků, vdov, strádajících spisovatelů, umělců, studentů a jiné. Skomírající Rakousko chtělo své síly osvěžiti sbírkami kovů, kaučuku, hadrů a jiného smetí, ale ani jediného skauta nedali jsme k této *činnosti válku podporující*, ač nejen mládež měšťanských a obecných škol s učitelstvem, ale i středoškolští profesoři byli ke sbírkám těm přímo komandováni. Ušli jsme také nechutnému *cvičením vojenským* a ani jediný *německý povel* skauti naši neuslyšeli... Také *časopis náš »Junák«* v tom směru se neprohřešil a hleděl si, pokud nebylo možno mluviti otevřeně, věcí speciálně skautských.

Neušli jsme sice přes to různým šikanám se strany úřadů, ale vyhnuli jsme se rozpuštění a stalo se to prostředky čestnými.

Pro nás vychovatele mládeže je ale ještě z jiného důvodu důležito, byla-li tato cesta správna. Bylo nutno obcházeti zákony, nařízení, vymykati se povinnosti ke státu před tváří mládeže. Nechovám obav, že by *vliv náš na mládež* býval nepříznivý. Význam mlčení a chování národa našeho za války byl jasný i dětem, tím spíše skautům a v táborech, při schůzkách a vycházkách našel každý vůdce příležitost promluviti jasně a ukázati, že stály tu proti sobě uměle vytvořené povinnosti k naší tak zv. vlasti a největší *povinnosti k Vlasti, Národu, Demokracii, Svobodě*, a tu pro každého čestného člověka byla jen jedna volba možnou. Provozujíce tuto činnost, již bych nazval velezrádou zázemí, udrželi jsme činnost oddílů v mezích slučitelných s přesvědčením vlastním a podařil se nám plán, zachovati skauting při životě, pracovati a nastřádati zkušeností pro dobu, kdy bude možno skauting organisovati *pro nás*.

Nyní, kdy možno každému všechno říci, co má na srdci, ozývají se mnozí, kteří v těch prokletých dobách ztratili také řeč a co více — stáli opodál se založenýma rukama, a teď vytýkají i těm, kdož byli veřejně činni, že nejednali a nemluvili tak radikálně.

Sabotage by Junák during World War I. (From Svojsík's speech on November 28, 1918.)

Document Text:

Introduction to the Annual Report of the Chief Leader, A. B. Svojsik
unanimously approved by the Committee and General Assembly of the association
"Junak-Czech Scout" on November 28, 1918.

First and foremost, before I provide a detailed account of our work in the past year, let me express a few introductory words:

Firstly, I am allowed to speak as a genuinely free individual, on the grounds of freedom, to my scout comrades - the citizens of the free Czechoslovak Republic. Here, I can speak openly.

It is truly gratifying to reminisce about the beginnings of our scouting journey. The leaders of our nation immediately recognized the enchanting power of scout education, and the vital role scouting could play in the development and, dare I say, the transformation of the Czech youth's character. Dr. Kramar warmly welcomed my initial articles and plans. When I presented Masaryk with the foundational work of "Scouting for Boys" in English, he greeted me with the words: "Ah, I am already familiar with that work, and I will support scouting here wherever I have influence." Čada, Drtina, Jirasek, Klofac, and other leaders among us not only showed a platonic interest in the movement but actively contributed to its cause.

However, the path was more challenging with the wider public. Criticisms, misunderstandings, attacks in the press - even personal ridicule and threats - were not uncommon. Yet, the core of scouting is so resilient that through diligent work, we have gradually gained ground and earned trust and respect, despite the narrow-mindedness of the Austrian bureaucratic mentality regarding youth education. Not long ago, we all suffered under its constraints. We introduced youth self-governance, club meetings, uniforms, life in nature, and many other practices that occasionally brushed against the restrictions imposed by school laws and disciplinary regulations. Some individuals in positions of authority quickly grasped the value of scouting. However, the Austrian system zealously guarded every youth movement to ensure it adhered to their prescribed mold, as they prepared to exert their dominance worldwide. It was against our will that the promising movement was stifled in its infancy, perceived as a threat.

In Vienna, uniformed generals and titled barons and counts paraded around, while in the midst of our determined work and the promising growth of our young organization, war broke out. Along with it came that well-known body of ticks that crushed and suffocated anything that breathed Czech.

Scouts who actively participated in public life and national campaigns could not simply remain indifferent when others, especially many of our own people, rushed to serve the "endangered homeland." We were left with two options: either cease all activities and start anew under unknown circumstances after years, or find an acceptable modus vivendi to preserve our honor and maintain at least an illusion of cooperation with the rest of the "Austrian" world. It was a form of silent resistance, a quiet revolution, which persisted even when parliamentarians and newspapers were silenced. In that sea of human misery that engulfed the world, the committee of "Junak" unanimously and in accordance with our ideals and scout promises decided to engage in humanitarian work. We organized actions to support wounded veterans, orphans, widows, struggling writers, artists, students, and others.

Austria, on the brink of collapse, sought to revitalize its strength through collecting metals, rubber, rags, and other scraps. Yet, we did not contribute a single scout to support this endeavor, despite not only the youth of bourgeois and public schools but even high school professors being directly commanded to participate in the collections. We also avoided repugnant exercises in the spirit of the Vienna regime, and our continued our activities with integrity. Our magazine, "Junak," also stayed true to its purpose, openly discussing matters specifically related to scouting whenever possible.

Though we faced various harassments from authorities, we managed to evade dissolution and remained true to our principles. Our actions were honorable.

For us, as educators of youth, it was crucial to consider whether this path was the right one. We had to navigate laws, regulations, and obligations to the state when facing the youth. I have no doubt that our influence on the youth would have been unfavorable had we acted differently. The significance of our nation's silence and behavior during the war was evident not only to children but also to scouts. In camps, meetings, and outings, every leader had the opportunity to speak clearly and demonstrate that we stood against the artificially imposed obligations to our former homeland and the holiest duty to our nation. To Democracy. To Freedom. For every honorable person, there was only one choice. While pursuing this activity, which some may label as treason behind the scenes, we maintained our troop activities within the bounds compatible with our own convictions. We accumulated experience for the time when scouting could be organized openly for us.

Now, when anyone can say everything, they have in their hearts, many who were also silent during those cursed times, or worse, stood idly with folded arms, are now criticizing those who were openly active, claiming they did not act or speak as radically. Sabotage by Junaks during the World War (From Svojsik's speech on November 28, 1918) - Page 1.

Czech plan for the establishment of a lodge from 1914. (From the notes of František Sísa.)

Document Text:

It was in the feverish and tumultuous time at the end of 1914, during the darkest oppression of the Czech nation when military absolutism held sovereign power over our bodies and souls as Czech citizens. It was during this time that the idea of establishing a Czech Masonic lodge in Prague was born. We began to organize underground revolutionary activities more firmly in our struggle for national liberation. We formed a network of trusted confidants to jeopardize and paralyze the functioning of the regime and provide intelligence services on military matters, war industries, the economic and financial state of the monarchy, railways, and all matters related to the war. It was then that the idea emerged that a Masonic lodge would be the most effective organization, both ideologically and practically, for achieving the goals we had set for ourselves—national liberation and cooperation with the Allied powers.

Ideologically, the basic doctrines and goals of freemasonry pointed the way. The Czech plan for the establishment of a lodge in 1911...

(From the notes of František Síse)

Authorization letter for František Sísa.

Document Text:

Cis. 12/23.
Translation of Attachment 4.
Orient Prague 25/1 923.
The Supreme Council of the 33rd and final degree of the Ancient and Accepted Scottish

Rite for Czechoslovakia appointed the Very Enlightened Brother František Sis as the 33rd delegate of the Supreme Council for France and dependent territories.

On behalf of the Supreme Council for Czechoslovakia,
Viktor Dvorský Alfons M. Mucha
Grand Chancellor General Secretary Sovereign Commander

Lodge card of the 18th degree issued by the Supreme Council in France.

Document Text:

Breve 18°.

In the name and under the auspices of the members of the Grand Inspectors General of the 33rd degree of the Ancient and Accepted Scottish Rite, who constitute the Supreme Council for France and dependent territories under the celestial vault with the zenith at 48°50' north latitude. Orient Paris. Equality, liberty, fraternity.

To our enlightened members, the Grand Inspectors General, princes of the royal secret, Knights Kadosh, Knights of the Rose Cross, the Great Elect, leaders and members of the Masons, all Masons of the Ancient and Accepted Scottish Rite spread across the earth, we announce that, considering the request submitted to us by our Very Enlightened Brother Grand Chancellor, and the favorable recommendation of Sovereign Chapter No. 521 Mozart in the Valley in Vienna, we, the Sovereign Grand Commander, Deputy Sovereign Grand Commander, members of the Grand Inspectors General of the 33rd and final degree of the Ancient and Accepted Scottish Rite, constituting the Supreme Council for France and dependent territories, have elevated and do elevate to the rank of Knight of the Rose Cross, the 18th degree of our rite, our dear Brother Mascha Ottokar, born on May 7, 1852, in Plzeň, from the regular Lodge No.... under the name "Eintracht" from the Orient of Vienna, registered under the general control of jurisdiction of our Supreme Council under number..., and desiring this Brother to enjoy the benefits, rights, privileges, and honors associated with this degree, to exercise its attributes in accordance with the constitutions, general and particular rules of our order, and to be recognized everywhere in this capacity, we have issued this duly confirmed breve, signed, sealed with the seals and stamps of our Supreme Council, to confirm both the degree conferred upon him and his registration in the special control of our 18th degree under number 97.

We request and command all regular Masons, all workshops within the jurisdiction and obedience of our Supreme Council, and we call upon all other Masons and regular Masonic lodges of all degrees in the name of the mutual obligations, friendship, and goodwill that bind and unite us to recognize our dear Brother Mascha Ottokar in the degree, title, and rank mentioned above, and to fulfill all the regulations prescribed by the laws of our order towards him.

Done and given at the meeting of the Supreme Council on the day of December 192?.
Signatures.

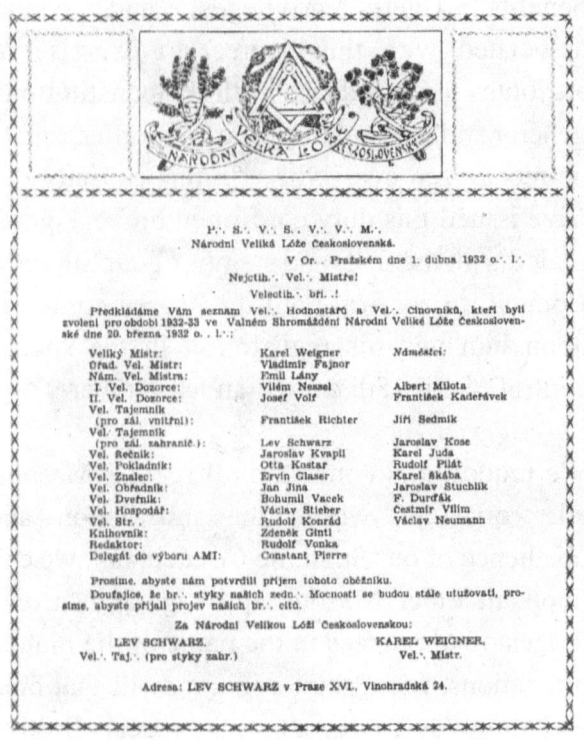

Charter of the Grand Officers of the National Grand Lodge of Czechoslovakia from 1982.

Document Text:

National Grand Lodge of Czechoslovakia.
In Prague, on April 1, 1932.
Most Worshipful Master!
Respected brethren!

We present to you the list of Grand Officers and Grand Officials who were elected for the term 1932-33 at the General Assembly of the National Grand Lodge of Czechoslovakia on March 20, 1932:

Grand Master: Karel Weigner Deputies:
Deputy Grand Master: Vladimir Fajnor Emil Lány
I. Grand Warden: Vilém Nessel Albert Milota
II. Grand Warden: Josef Volf František Kadeřávek
Grand Secretary (for internal affairs): František Richter Jiří Sedmik
Grand Secretary (for foreign affairs): Lev Schwarz Jaroslav Kose
Grand Orator: Jaroslav Kvapil Karel Juda
Grand Treasurer: Otta Kostar Rudolf Pilát
Grand Expert: Ervin Glaser Karel Škába
Grand Ritualist: Jan Jína Jaroslav Stuchlík
Grand Tyler: Bohumil Vacek F. Durďák
Grand Steward: Václav Stieber Čestmír Vilím
Grand Librarian: Zdeněk Gintl
Editor: Rudolf Vonka
Delegate to the AMI Committee: Constant Pierre

Please confirm the receipt of this circular.

Hoping that the fraternal connections between our Masonic Powers will continue to strengthen, we request you to accept the expression of our brethren's sentiments.

For the National Grand Lodge of Czechoslovakia:

LEV SCHWARZ, KAREL WEIGNER,

Grand Secretary (for foreign affairs) Most Worshipful Master.

Address: LEV SCHWARZ in Prague XVI, Vinohradská 24.

List of Grand Officers of the National Grand Lodge of Czechoslovakia from 1932

Emblem of the 30th degree: Knight Kadosh. (Skulls with a royal crown, civic laurel wreath, and papal tiara symbolize the powers rejected by Freemasonry. The dagger hanging from the sash of the 30th degree symbolizes the idea of revenge.)

Chamber of Contemplation that every initiate must go through.

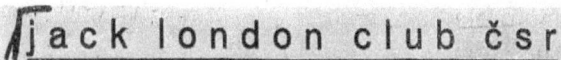

Letter protesting the participation of Sokol in the Berlin Olympics.

Document Text:

Dear Participants,

We would like to inform you about the outcome of our Friday discussion on the participation of Sokol members in the Berlin Olympics. Representatives of ÖAAÜ, Hradní Labor, Ilbk&bi, and the Student Youth Front attended the meeting. It is evident that many people view the Olympic Games with great enthusiasm but deeply regret that they are being held in Berlin. It is more than likely that these games will serve to strengthen and promote the inhumane regime in Europe.

This was the main reason why the discussion resulted in a stance against Sokol's participation in Berlin. We, the progressive young generation, cannot support an event organized by undemocratic organizations and held in a country that poses a threat to the personal safety of athletes and participants. It goes against our fundamental principles and the very essence of democracy, which upholds the basic rights of individuals.

The meeting proposed organizing a broad democratic plebiscite among the members of the organization, allowing their voices to be heard and serving as a guideline for the decision-making process of Sokol. This plebiscite would reflect the wishes of the wider

Sokol community and provide valuable input for the leadership of the organization.

Our concern regarding the Sokol's participation is not exhausted by this letter, and we will continue to closely monitor the developments. We hope that this letter contributes to your decision-making process and aligns with the desires of the Czechoslovakian public.

With utmost respect,

Z. (illegible)-
Jack London Club
Letter protesting Sokol's participation in the Berlin Olympics

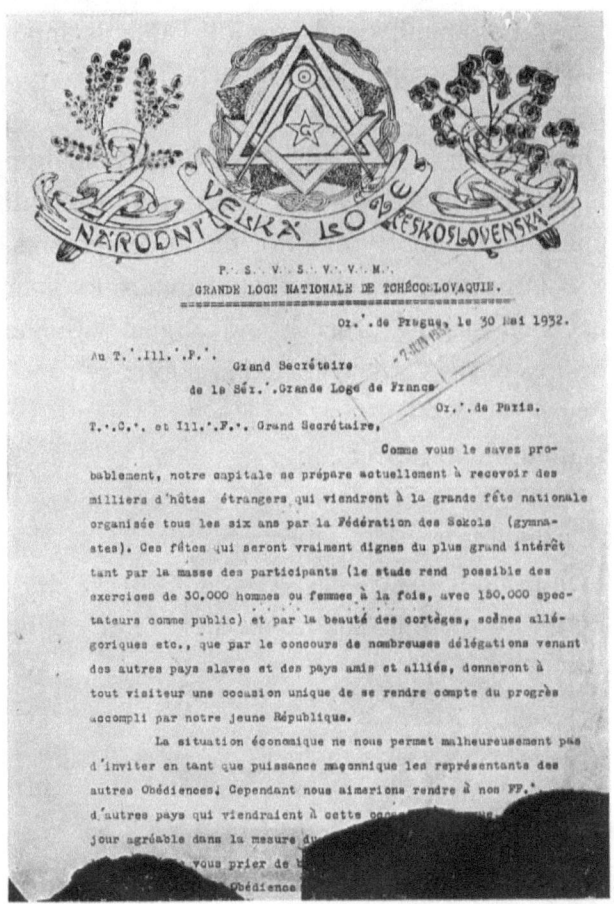

Part of a letter to the Grand Lodge of France inviting them to the Sokol gathering in 1982.

Document Text:

[on next page]

> adresse. Nous vous prions en même temps de leur indiquer notre
> adresse pour qu'ils puissent s'adresser à nous pour toute sorte
> d'aide dont ils auraient besoin. Nous pourions dans la mesure de
> nos possibilités nous charger sur leur demande de leur faire retenir des chambres d'hôtel, des billets d'entrée etc. pourvu qu'on
> nous informe à temps i.e. au plus tard jusqu'au 15 juin prochain
> (les jours des grandes fêtes dont je vous envoie un petit programme
> préliminaire, étant les 2,3,4,et 5 juillet prochain).
>
> Nous aurions beaucoup de plaisir si nous pouvions saluer à
> Prague à cette occasion des FF∴ de votre Obédience et de reserrer
> par leur intermédiaire les liens fraternels qui nous unissent déjà
> avec votre pays et avec votre francmaçonnerie.
>
> En attendant le plaisir de vous lire prochainement je vous
> prie de croire, T∴C∴ et Ill∴F∴ Grand Secrétaire, à l'assurance de mes sentiments les plus fraternels.
>
> L. Schwarz
> Grand Secrétaire (aff.ét.)
> de la Grande Loge
> Nationale de Tchécoslovaquie
>
> Adresse pour lettres:
> L. Schwarz
> à Prague-Smíchov
> Vinohradská No. 24
> Tchécoslovaquie.

Part of a letter to the Grand Lodge of France inviting them to the Sokol gathering in 1932 (back side).

Document Text:

Orient Prague, May 30, 1932.

To the highly enlightened Brother Grand Secretary of the noble Grand Lodge of France, Orient Paris.

Dear and enlightened Brother Grand Secretary, as you may already know, our capital is now preparing to welcome thousands of foreign guests who will come to attend the grand national festivities organized every six years by the Sokol Community (gymnasts). These festivities truly deserve the greatest attention, both due to the number of participants (the stadium allows simultaneous exercises of 30,000 men or women with the presence of 150,000 spectators) and the beauty of the processions, allegorical scenes, etc. Moreover, many delegations from other Slavic countries and from friendly and allied nations will also be present. This will provide each visitor with a unique opportunity to get an impression of the progress achieved by our young republic.

Unfortunately, the economic situation does not allow us, as a Masonic authority, to invite representatives of other obediences. However, we would like to make the stay of our Brothers from other countries who come to Prague on this occasion more enjoyable (text disrupted...). We kindly request you to inform them of our address so that they can approach us for any assistance they may need. Within our capabilities, we could provide them with assistance such as arranging hotel rooms, tickets, etc., provided we are informed in a timely manner, no later than June 15 (the days of the grand festivities, for which I am sending you a preliminary program, are July 2nd, 3rd, 4th, and 5th).

It would be a great pleasure for us to greet brethren from your obedience in Prague on this occasion and to strengthen the fraternal bonds that already exist between our countries and our Freemasonry through their presence.

While looking forward to hearing from you soon, please accept, dear T.-.C.-. and M.-.W.-.F.-. Grand Secretary, the assurance of my most fraternal sentiments.

Grand Secretary (affiliated)
of the Grand National Lodge of Czechoslovakia
Address for correspondence:

Excerpt from the letter of the Grand Lodge of France inviting foreign Masonic brethren to the Sokol festival in 1932.

Binding of the book "Skaut-malý zednář" (Scout-Little Freemason).

Title page of the book "Skaut-malý zednář" (Scout-Little Freemason).

Document Text:

SCOUTT - SMALL MASON
AND WITH THE KEY IT IS RATHER NOT
ZDENKO STERI

H. D. E. L. L. T. Y. O. M. O.
Title page of the "Skaut-malý zednář" document

dením nováčka a jeho přípravou k nováčkovské zkoušce.

I sestry mají skauti - skautky, ale ty nesmí se súčastňovati společných prací - táboření ani klubovního života - s hochy. Jen jednou do roka bývá společný, i rodičům a příznivcům přístupný společný slavnostní táborový oheň.

K sestrám skautkám chovají se skauti s uctivou rytířskostí.

Slunce, měsíc a hvězdy byly kultem snad prvního člověka, jemuž mozek dal schopnost pozorovati a usuzovati. Byly kultem starých Číňanů, Mexičanů, Asyřanů, Egypťanů, a jsou dodnes kultem skautstva, a to jak v přírodě, tak v životě. V životě byl naším sluncem ideál svobodného zednářství president Osvobo-

Page from the secret book "Skaut-malý zednář" (Scout-Little Freemason).

Document Text:

The new boy and his preparation for the new boy examination.

Sisters have scouts - girl scouts, but they are not allowed to participate in common activities - camping and club life - with the boys.

Only once a year, there is a common, open campfire for parents and supporters.

Scouts treat the scout sisters with respectful chivalry.

The sun, moon, and stars have been worshiped by perhaps the first human whose brain gave him the ability to observe and reason. They were worshiped by ancient Chinese, Mexicans, Assyrians, Egyptians, and are still worshiped by scouts today, both in nature and in life.

In life, our sun was the ideal of free masonry, President Osvobo – ...

ditel Dr. T. G. Masaryk, protektor a veliký příznivec Svazu Junáků skautů a skautek R. Č. S.

Naším měsícem jest nynější president Republiky Dr. Edvard Beneš, starosta Svazu skautů až do své volby presidentem, od té doby čestný starosta Svazu. Nyní jest druhý president Republiky naším druhým protektorem.

Mezi hvězdami jest nám polárkou zesnulý prof. Dr. Karel Weigner, který napsal knížku „Význam skautingu v tělesné výchově."

Vy, velicí zednáři zašlých věků, shlédněte s Věčného Orientu na naši mládež; snad potěší Vás vzrůst sémě, Vámi v lidstvo zasetého!

A ctihodní mistři dnešních dob, Svou

Page from the secret book "Skaut-malý zednář" (Scout-Little Freemason).

Document Text:

Dedicated to Dr. T. G. Masaryk, protector and great supporter of the Scout Union of Scouts and Girl Scouts of Czechoslovakia.

Our moon is the current President of the Republic, Dr. Edvard Beneš, former Mayor of the Scout Union until his election as President, and now the Honorary Mayor of the Union. He is now our second protector as the second President of the Republic.

Among the stars, we have the late Professor Dr. Karel Weigner as our guiding star. He wrote the book "The Significance of Scouting in Physical Education."

You, revered masons of bygone eras, look down from the sacred Orient upon our youth; perhaps the growth of the seed sown by humanity will please you!

And to the respected masters of today, in your wisdom and guidance, may our young scouts find inspiration and knowledge.

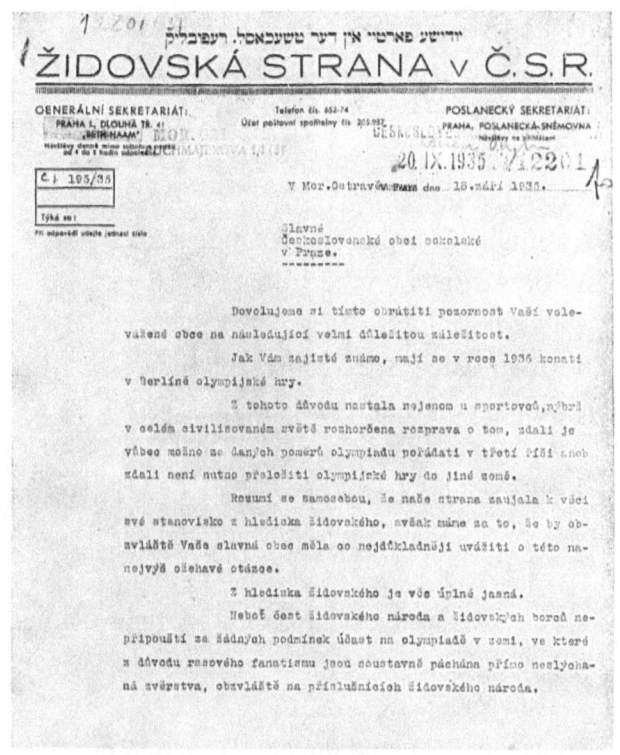

Jewish dedication to Sokol.

Document Text:

JEWISH PARTY in CZECHOSLOVAKIA

GENERAL SECRETARIAT:
PRAGUE I., DLOUHÁ TŘ. 41

T.Won Ei». 652-74

PARLIAMENTARY SECRETARIAT:
Account Postal No. 205,937. PRAGUE, REPUBLIC SQUARE

Date: 195/35
Replying to your communication dated:
In Mor. Ostrava on the 15th of April 1935...

To the esteemed representatives of the municipal community of Prague.

We take this opportunity to draw your attention to the following very important matter.

As you are surely aware, the Olympic Games are scheduled to take place in Berlin in 1936.

Due to this, a heated debate has arisen not only among athletes but also in the civilized world at large, regarding whether it is possible to hold the Olympics under the current circumstances in the Third Reich, or whether it is necessary to relocate the Olympic Games to another country.

Naturally, our party has taken its stance on this matter from a Jewish perspective. However, we believe that your illustrious community should carefully consider this highly sensitive issue.

From a Jewish perspective, the matter is clear. The honor of the Jewish nation and Jewish athletes does not permit their participation in the Olympics in a country where unspeakable atrocities are systematically committed against the Jewish people due to racial fanaticism.

Jewish Regards to Sokol.

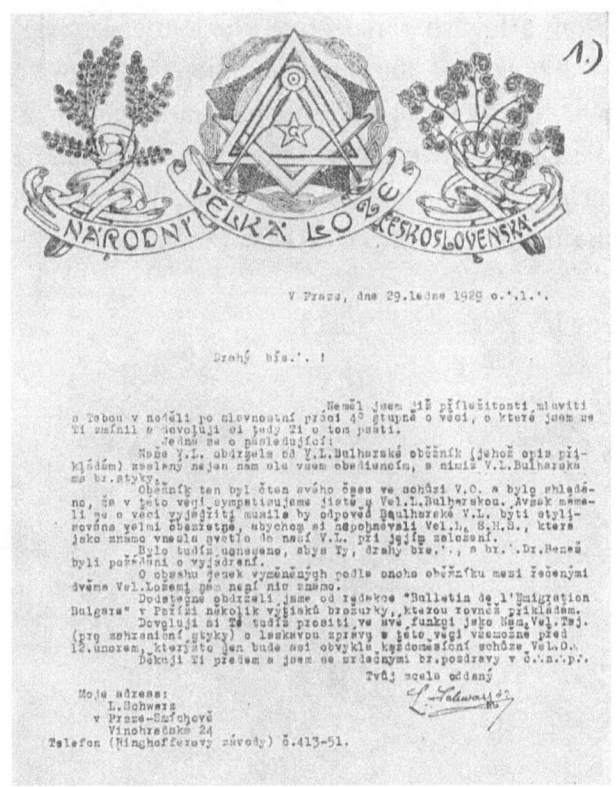

Letter to Dr. Kamil Kroft from the National Grand Lodge of Czechoslovakia.

Document Text:

Dear Dr. Kamil Kroft,

I hope this letter finds you well. I am writing to bring your attention to an important matter that we have previously discussed and to request your assistance in addressing it.

The issue at hand concerns the recent circular sent by the Grand Lodge of Bulgaria (a copy of which is enclosed) to not only our Lodge but to all other Lodges with which the Grand Lodge of Bulgaria maintains relations. This circular was discussed during a meeting of our Lodge, and it was determined that there are certain concerns regarding the content. However, if we were to express our views on the matter, it would require a careful approach so as not to forget the enlightening guidance provided by our Grand Lodge when it was established.

Therefore, it was decided that you, dear Brother, along with Brother Benes, should be approached to provide your opinions on the matter. Specifically, we would like to hear your thoughts on the exchange of the mentioned documents between the two Grand Lodges, as we currently have no further information on this matter.

Additionally, we have received several copies of the brochure "Bulletin de la Logeation Bulgare" from the editor in Paris (also enclosed). In light of these developments, I kindly request that you provide any information or updates on this matter before February 12th, which is when our regular monthly meeting of the Grand Lodge is scheduled to take place.

Thank you in advance for your assistance, and I send you my warm fraternal greetings.

Yours faithfully,

L. Schwarz
Address:
Vinohradská 24, Prague-Smíchov
Telephone: C.413-51

Letter from the National Grand Lodge of Czechoslovakia to Dr. Kamil Kroft.

Letter from Freemason A. Mucha, 33rd degree, indicating that Jan Masaryk was a Freemason.

Document Text:

ORDO 48 CHA.0
N E J V Y Š Š I RADA

33TÉHO A POSLEDNÍHO stupné ŘADU
ŠKOTSKÉHO STARCA
PRO ČESKOSLOVENSKO

[On next page]

Letter from Freemason A. Mucha, 33rd degree, indicating that Jan Masaryk was a Freemason (reverse side).

Document Text:

Dear and highly esteemed Brother,

I would like to extend my heartfelt gratitude to you for the kindness with which you provided me with the address of Mr. Ivanhoe Rambosson. I have used it to write to him, and he has already responded. I have known him for a long time, and he has always been very dear to me, and continues to be so.

Today, my letter will surely test your fraternal patience, I am afraid to say. However, you must know what it is about:

On the 26th of this month, we will have a solemn ceremony in our Lodge, actually in our Freemasonry as a whole, to inaugurate the National Grand Lodge of Czechoslovakia. A delegation from the Grand Lodge of Yugoslavia in Belgrade, led by the Sovereign Grand Commander, will arrive to bring us the light.

We would like to present ourselves in the appropriate manner during this ceremony, but several of our Brothers of the 33rd degree do not have the insignia of their rank, namely the silver suspended eagles and the crosses of the 33rd degree.

In our predicament, I therefore ask you if you could entrust someone from your vicinity to purchase the insignia for us, which may be available for sale in a store whose address you surely know.

We require insignia for 8 of our Brothers, specifically:

8 silver suspended eagles,
8 crosses of the 33rd degree.

However, due to the extremely limited time (unfortunately, we could not find out earlier how many of these Brothers there would be), we would be immensely grateful if you could send us this package by airmail, which delivered your shipment to us in just one day (within six hours) - still in time.

Please include an invoice for the insignia and the courier, and we will promptly send you the money through the postal service as a token of our gratitude.

Would it be possible to call our Brother, Legation Councillor Jan Masaryk, the son of our President, who happens to be in Paris these days - he would gladly provide you with the funds for the requested purchase.

In my next letter, I will write to you about our situation, which remains very favorable, and our Masonic plans. For now, please accept, dear and enlightened Brother, my assurance of fraternal devotion.

Mucha/33.

Grande Loge de Bulgarie.

Or. de Sofia, le 24 Octobre 1928

No.1462.

Tr∴Ill∴Gr∴M∴
Tr∴chers FF∴,

Nous avons attendu de longs mois avant de répondre à la planche de la Grande Loge de Yougoslavie, par laquelle celle-ci se montrait si injustement injurieuse envers la Grande Loge de Bulgarie.

Nous ne voulions pourtant pas et nous ne voulons pas encore envenimer le débat. Nous ferons seulement en sorte cette fois de faire remettre notre réponse à Belgrade par les moyens les plus sûrs, afin de ne plus nous exposer au reproche si injuste qui nous fut fait de ne pas lui avoir adressé notre circulaire en faveur de nos compatriotes macédoniens.

Cette circulaire ne visait pas à une révision du statut territorial de la Macédoine. C'eût été de notre part une intrusion dans la politique, profane que nous interdissaient à la fois notre respect des traités et notre obéissance à nos vénérables traditions.

Nous n'avons été inspirés que par le désir de voir appliquer aux habitants bulgares de cette contrée les principes de justice compatibles avec leur dignité de citoyens. Nous demandions que la Franc-maçonnerie Universelle voulût bien intervenir pour que les Macédoniens demeurés dans leur petite patrie, à laquelle ils se trouvent aujourd'hui incorporés, qu'ils puissent suivre les exercices de leur culte, obéir à leurs coutumes ethniques et qu'il ne leur fut pas imputé comme crime de lire, de parler ou d'écrire la langue de leurs aïeux.

Nous signalions à la conscience maçonnique les exactions dont ils sont les victimes, et loin de nous attendre à la protestation de la maçonnerie yougoslave, nous osions même espérer qu'elle userait de son influence pour tenter de les faire cesser.

On nous a répondu à Belgrade par des paroles violentes. Nous n'emploierons pas la même méthode de discussion.

Nous demandons respectueusement à l'Association Maçonnique Internationale de vouloir bien - ainsi que la Société des Nations le fait pour les contestations entre les peuples - nommer des enquêteurs qui jugeront, en toute impartialité, si la cause que nous défendons n'est pas une cause humaine, digne d'attirer l'attention, la sympathie et enfin la noble assistance de tous nos frères.

On a outragé également notre premier Grand Maître, le frère Général Prothoghgroff, maintenant passé déjà à L'Or∴Eternel, sur la vie duquel une légende abominable a été répandue. La Francmaçonnerie dont le concours est acquis à la réparation de toutes les injustices, mêmes profanes, voudra bien sur ce point encore ordonner une enquête.

Et ainsi nous pourrons espérer, que la lumière apportée sur ces faits douloureux dissipera tous les malentendus et fera enfin régner l'entente de la Grande Loge de Bulgarie avec la Grande Loge Yougoslave - entente que nous désirons ici de tout notre cœur de maçons.

Veuillez agréer, très Ill∴Gr∴M∴ et très chers FF∴, l'expression de nos sentiments frat∴ dévoués.

Le Gr∴M∴ P.Midileff

Le Gr∴Secr∴G.N.Koledaroff.

Report from the National Grand Lodge of Bulgaria (appendix to Kroft's letter dated January 29, 1929).

Document Text:

Ref: No. 1462.

GRAND LODGE OF BULGARIA.
Orient Sofia, 2nd October 1928.

Very enlightened Grand Master, dear Brothers,

We have been waiting for many months to respond to the circular of the Grand Lodge of Yugoslavia, which unjustly and offensively addressed the Grand Lodge of Bulgaria.

However, we did not want, and do not want even today, to engage in a debate. This time, we will only take care to ensure that our response is safely delivered in Belgrade, so as not to expose ourselves to such an unjust accusation made against us for not sending our circular, in which we supported our Macedonian compatriots.

This circular did not concern the revision of the territorial status of Macedonia. That would have been an interference in secular politics, which is prohibited by our respect for treaties and our obedience to venerable traditions.

We were led solely by the desire for the principles of justice, which are commensurate with the dignity of the Bulgarian population in this region. We requested the Universal

Freemasonry to kindly intervene so that they would regard the Macedonians who remained in their homeland, now attached to their larger homeland, without distrust, allowing them to attend their religious services, adhere to their national customs, and not consider it a crime for them to read, speak, or write in the language of their ancestors.

We presented to the Masonic conscience a report on the unjust taxation imposed upon them, and we did not expect a protest from Yugoslav Freemasonry; in fact, we dared to hope that they would use their influence to try to put an end to it.

In Belgrade, they responded with strong words. We will not employ the same methods in the debate.

Respectfully, we request the International Masonic Association, as the League of Nations does in international disputes, to appoint an investigative commission that will impartially assess whether the issue we are defending is a human matter deserving attention, sympathy, and ultimately the noble assistance of all our brethren.

Our first Grand Master, Brother General Protogerov, who has already departed to the Eternal Orient, was also offended by the spread of an abhorrent legend about his life. Freemasonry, whose aid is secured to rectify all injustices, including secular ones, kindly orders an investigation into this matter as well.

Thus, we can hope that the clarification of these painful facts will dispel all misunderstandings and restore harmony between the Grand Lodge of Bulgaria and the Grand Lodge of Yugoslavia, a harmony we wholeheartedly desire from our Masonic hearts.

Please accept, very enlightened Grand Master and dear brethren, this expression of our fraternally devoted sentiments.

Grand Secretary (Seal)
Brother K. Midtleff

Grand Orator
Brother J. Koláauioff.

Report of the Grand Lodge of Bulgaria
(attached to the letter to Kroftovi dated January 29, 1929)

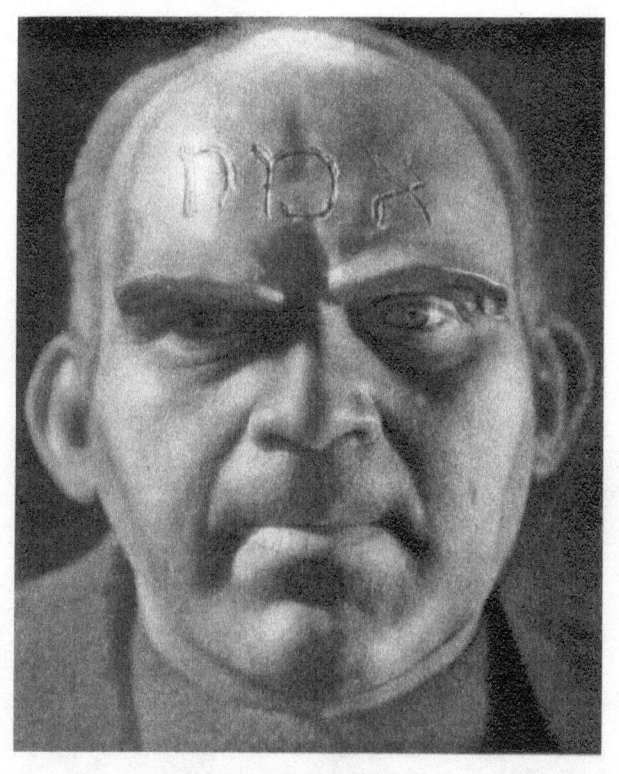

Mask of the Golem, the main character from the film of the same name[vi] *filmed at Barrandov Studios in Prague.*

Section of the Prague Jewish Cemetery.

Monument with the symbol of the Karpeles family at the Prague Jewish Cemetery.

Former Minister of Education, Univ. Professor Dr. Jan Kapras, Freemason of the 33rd degree and the last representative of the Grand Commander of the Supreme Council.

About the Author

The author, Walter Jacobi (July 2, 1909- May 3, 1947), was born in Munich, Germany. He was a German lawyer. He joined the Nazi Party during his studies at Martin Luther University in the early 1930s and served as a soldier in the Nazi SA unit. He became a SS officer, holding the rank of SS-Obersturmbannführer (Lieutenant Colonel).

In 1938, Jacobi was assigned to the headquarters (Zentralabteilung II) in Berlin where several significant events took place regarding the Sudeten Germans. Following the annexation of Austria to the German Empire, Jacobi played a role in the coordination and administration of German policies towards the Sudeten Germans. This included organizing propaganda campaigns, facilitating the integration of Sudeten Germans into Nazi organizations, and exerting influence to advance German interests in the region. These activities aimed to promote German nationalism and strengthen the Nazi regime's control over the Sudetenland, ultimately leading to the Munich Agreement and the subsequent German occupation of the area.

He was assigned to Prague just five days after the Nazi occupation on March 20, 1939, where he became notorious as the head of the Sicherheitsdienst (SD).

In September 1939, Jacobi played a significant role in implementing Nazi policies and he callously approved the persecution and execution of Czech students and intellectuals who protested against the university's closure. As the head of the Sicherheitsdienst (SD), Walter Jacobi collaborated closely with Reichsprotektor Konstantin von Neurath, the

Nazi-appointed representative overseeing the Protectorate of Bohemia and Moravia.

After the Nazi attack on Poland, the German authorities intensified their repression, leading to heightened tensions in Prague. On October 28, 1939, clashes erupted, resulting in the fatal injury of a worker named Václav Sedláček and a medical student named Jan Opletal. Opletal's funeral triggered further demonstrations and provoked German retaliation. Walter Jacobi, already involved in the events, played a role in compiling lists of student leaders from his files. Gestapo received orders to locate and arrest them.

Together, they orchestrated the infamous Sonderaktion Prag on November 17, 1939. This brutal operation involved the closure of all Czech universities and colleges, the arrest of 1,850 students, and the execution of nine student leaders, including František Skorkovský. Additionally, more than 1,200 Czech students were interned in the Sachsenhausen concentration camp.
Subsequently, nine student leaders from Jacobi's list were executed at the Ruzyně barracks. These actions were part of a systematic campaign to suppress Czech

resistance and impose Nazi control over the occupied territory.

During the war, he issued immediate orders for the execution of those opposing the German regime and took part in the massacre of the village of Lidice near Kaldeno. In the massacre of Lidice, approximately 173 men were executed, 184 women were sent to concentration camps, and 88 children were taken from their families and subjected to various forms of abuse and exploitation. The village itself was completely destroyed, with homes demolished and the land plowed over. The total death toll, including those directly killed and those who died later due to the repercussions of the massacre, is estimated to be around 340 individuals.

In 1942, Walter Jacobi ascended from a department head to become the leader of the SD office in the Protectorate. During this time, he gained recognition not only for his intelligence work but also for his involvement in propaganda, including the publication titled "Golem, the Broom of the Czechs." The book specifically targeted Czech Freemasons, with a particular focus on Jan Kapras, the Minister of Education and a native of Brno, who was known for his collaboration with the Czech resistance. According to Jacobi, their investigation revealed that Kapras had attained the 33rd degree within the Freemasonry organization, as evidenced by the documents they uncovered.

Jacobi's attack and the activities of the SD had far-reaching consequences, leading to the downfall of Jan Kapras and instigating changes within the government of the Protectorate. The SD and its operatives extended their network throughout the country, infiltrating various sectors including politics, education, media, industry, and social organizations. With 16 rural offices across the Protectorate, the SD diligently gathered information and compiled comprehensive reports, which were closely guarded and shared only among a select group of insiders. These reports covered a wide range of topics, including the prevailing sentiments in key armament factories like the Škoda Works in Pilsen and the agricultural regions, providing valuable insights into the state of affairs in these crucial sectors.

Although not as widely recognized within the Nazi security apparatus, Walter Jacobi held considerable power and influence. His role in the Protectorate of Bohemia and Moravia made him a figure of significant authority, often referred to as the "uncrowned king."

Following the end of World War II, Walter Jacobi experienced a period of relative luck compared to some other prominent Nazi figures. He successfully evaded capture and found employment at an American airport in Schleißheim, Bavaria, in May 1945. However, his fortune changed when he was identified by American investigators in mid-September 1945. Based on a Czechoslovak arrest warrant, he was apprehended and transported from Plzeň to Prague. Subsequently, as a convicted war criminal, Jacobi faced trial before the Extraordinary People's Court in Prague, where he was found guilty and sentenced to death.

On May 3, 1947, at the Prague-Pankrac prison, he was executed by hanging.

About the Translator

Kytka Hilmarová, a Prague native and political refugee, embarked on a transformative journey at a young age when she and her parents sought asylum in the United States in 1968. As an accomplished author, translator, and publisher, Hilmarová has left an indelible mark on the literary world, bridging the gap between Czech literature and English-speaking readers.

With over 200 books brought to life as a prolific ghostwriter and a portfolio of translating more than 100 Czech literary works into English, Hilmarová acts as a vital bridge connecting Czech literature with a global audience. Her visionary approach and unwavering commitment to preserving and promoting Czech culture, history, tradition, and literature have ensured that the legacy of Czech literary works remains alive, vibrant, and cherished for generations to come.

As the founder of Czech Revival Publishing, Hilmarová showcases the rich tapestry of Czech literary gems, fostering cultural exchange and expanding the global reach of Czech authors. Through her captivating works and translations, she invites readers to immerse themselves in the enchanting world of Czech literature, offering a glimpse into its diverse themes, profound emotions, and timeless wisdom.

Join Kytka Hilmarová on a literary journey that illuminates the treasures of Czech literature, history, and tradition. Her exceptional talent, resilience, and relentless pursuit of bridging cultures make her an indispensable figure in bringing the richness of Czech literature to English-speaking audiences, ensuring its enduring legacy for years to come.

10% of book proceeds support the preservation of Czech culture in the United States.

Translator's Endnotes

[i] **Maffie** served as the principal organization of the Czech domestic resistance during World War I, overseeing intelligence operations, conspiracy activities, information dissemination, and maintaining connections with the foreign section. More than 200 individuals participated in its activities. The name "Maffie" was chosen in admiration of the Sicilian organization known as the Mafia.

The history of Maffie began with the formation of a domestic committee in December 1914, following Professor T.G. Masaryk's departure abroad. From the outset, Maffie aimed to support T.G. Masaryk's foreign actions. It held consistently anti-Austrian and pro-Entente positions. In March 1915, its executive board was established, consisting of Edvard Beneš, Karel Kramář, Alois Rašín, Josef Scheiner, and Přemysl Šámal.

In September 1915, following Edvard Beneš's forced emigration, Přemysl Šámal assumed leadership of Maffie. During this year, Karel Kramář was arrested. Josef Scheiner, the mayor of the Czech Sokol Community, was among the founding members of Maffie and was arrested on May 21, 1915. He was escorted with Kramář to investigative detention in Vienna. The reasons for Scheiner's arrest included his involvement in Slavophile organizations, his contacts with Slavic countries and the United States, as well as an affair involving Sokol identification papers found in possession of Sokol members on the Russian front.

Scheiner was released on July 21, 1915, due to lack of evidence. His accusation was also one of the reasons for the suspension of activities of the Czech Sokol Community on November 24, 1915. Kramář was sentenced by an Austrian court in 1916 to death for treason and espionage. In November 1916, Emperor Franz Joseph I commuted his sentence to life imprisonment. After his death, Crown Prince Charles I, who pursued a friendly policy towards Slavs, granted amnesty to all those convicted in this trial, and in 1917, Kramář was released. Another member of Maffie, the journalist and economist Alois Rašín, was also tried in the same trial.

Members of the Maffie after receiving revolutionary medals at Prague Castle

Throughout World War I, Maffie's activities were interconnected with the activities of the Czechoslovak National Council (established in 1916 in Paris) and, toward the end of the war, with the activities of the Czechoslovak National Committee formed in June 1918. To facilitate the transmission of important messages, Maffie established the Alarm network, consisting of telephone operators, cyclists, and runners. Secret couriers played a crucial role in connecting the domestic resistance with the resistance abroad. One notable courier was the renowned opera singer Ema Destinnová. In September 1918, Maffie prepared for a non-violent military coup and the subsequent assumption of power.

[ii] **"Chrám mládeže"** (The Temple of Youth) was a concept within the Sokol movement that represented the ideal vision of a physical and moral educational space for the

youth. It aimed to provide a holistic environment where young people could develop their physical, intellectual, and moral capacities. The idea of the "Chrám mládeže" emphasized the importance of nurturing the youth through physical exercise, education, and character development.

The concept of the "Chrám mládeže" reflected the Sokol movement's belief in the transformative power of physical education and its commitment to shaping well-rounded individuals. The Sokol gymnastic halls, where physical training took place, were seen as more than just exercise spaces. They were envisioned as places of inspiration and learning, fostering discipline, camaraderie, and the pursuit of excellence.

The "Chrám mládeže" concept also emphasized the moral and ethical aspects of youth education. It aimed to instill values such as patriotism, selflessness, and dedication to the nation. The Sokol movement believed that by cultivating physical fitness, intellectual development, and moral character in the youth, they could contribute to the betterment of society as a whole.

The concept of the "Chrám mládeže" encapsulated the Sokol movement's holistic approach to youth education, emphasizing physical, intellectual, and moral development. It symbolized the ideal environment where young people could grow into active, responsible, and morally upright citizens.

[iii] **"Mal du siècle"** is a French term that translates to "the malady of the century" or "sickness of the age." It refers

to a cultural and emotional condition prevalent in the 19th century, particularly during the Romantic period. It reflects a sense of melancholy, disillusionment, and dissatisfaction with the state of society and the human condition.

The mal du siècle was often associated with feelings of alienation, existential angst, and a longing for something beyond the mundane realities of life. It was a response to the rapid social, political, and industrial changes taking place during that time, including urbanization, the rise of capitalism, and the impact of the Industrial Revolution.

Artistic and literary figures of the period, such as poets and writers, often expressed this sentiment through their works, emphasizing themes of individualism, introspection, and a search for meaning and transcendence. The mal du siècle reflects a romanticized view of suffering and an idealization of the past, longing for a lost era of purity, simplicity, and emotional intensity.

[iv] **In 1938, the Czech Scout movement, known as "Junák,"** was an active and influential organization in Czechoslovakia. Junák was founded in 1914 as "Junák - Czech Scout" and had been growing steadily in the years leading up to 1938. It had become the largest youth movement in Czechoslovakia, with a membership of around 40,000 boys and girls.

Under the leadership of Dr. Zdenko Štekl, the head of Junák, the organization aimed to provide a new form of education and upbringing for Czech youth. Junák emphasized discipline, outdoor activities, and the development of physical and mental abilities. The movement promoted the ideals of scouting, including self-reliance, teamwork, and service to others.

During this time, Junák had close connections with the international scouting movement, particularly with British and American scouts. The organization sought to maintain strong ties with scouts from other countries and had participated in international jamborees (large youth camps) to foster international cooperation and friendship.

The Czech Scout movement had prominent supporters, including President Tomáš Garrigue Masaryk and President Edvard Beneš, who served as the protectors of Junák. Masaryk, in particular, was known for his support of scouting and saw it as a means of promoting youth development and international understanding.

While the political climate in Czechoslovakia was becoming increasingly uncertain in 1938 due to rising tensions with Nazi Germany, Junák remained committed to its ideals and continued its activities. The organization maintained its focus on scouting principles, outdoor adventures, and community service.

It's important to note that during this time, the Czech Scout movement also had connections to Freemasonry. Some leaders within Junák, such as Professor Karel Weigner, were known Freemasons. These connections were reflected in certain symbols and rituals within the organization.

ᵛ **It's important to clarify that the statements provided reflect an anti-Semitic and conspiracy theory narrative that promotes harmful stereotypes and falsehoods.** Conspiracy theories, such as those linking Freemasonry and Jewish people to world domination, have been used historically to promote prejudice, discrimination, and hatred.

It is not accurate or fair to suggest that Jewish people, as a whole, are engaged in a conspiracy to control European state and national life or that they have achieved their positions through Freemasonry. Such claims are baseless and have been thoroughly debunked by reputable sources.

In today's world, it is crucial to foster understanding, respect, and tolerance among different communities and combat prejudice and discrimination. It is essential to approach discussions about religious, ethnic, and cultural groups with empathy, relying on accurate information and avoiding stereotypes or conspiracy theories.

ᵛⁱ **Le Golem (Czech: Golem)** is a 1936 Czechoslovakian monster film directed by Julien Duvivier in French. Set in a Prague ghetto, the story revolves around the oppression of poor Jews by the Holy Roman Emperor, Rudolf II. The Jews consider awakening the Golem, a creature kept in Rabbi Jacob's attic. When Countess Strada, Rudolph's mistress, is saved during a food riot, she convinces De Trignac to steal the Golem. The Golem's disappearance leads to Rabbi Jacob's arrest and the threat of hanging for the Jews involved. Rachel seeks De Trignac's help to rescue Jacob, and the Golem wreaks havoc in the palace before ultimately disintegrating. The film's screenplay faced legal issues, and it was shot at Barrandov Studios in Prague.

The cast includes notable actors such as Harry Baur as Rudolph II and Ferdinand Hart as the Golem. The film premiered in Paris and was a sequel to Paul Wegener's 1920 film "The Golem: How He Came into the World." It was released in Britain with a shorter running time and later reissued as "The Legend of Prague." Overall, the film received moderate praise, with critics appreciating the acceptance of fantastic elements and Harry Baur's performance as the emperor.

www.ingramcontent.com/pod-product-compliance
Lightning Source LLC
Chambersburg PA
CBHW020945230426
43666CB00005B/178